AF361928

Lightning Modeling and Its Effects on Electric Infrastructures

Lightning Modeling and Its Effects on Electric Infrastructures

Editors

Massimo Brignone
Daniele Mestriner

MDPI • Basel • Beijing • Wuhan • Barcelona • Belgrade • Manchester • Tokyo • Cluj • Tianjin

Editors

Massimo Brignone
Naval and Electrical
Engineering Department,
University of Genoa
Italy

Daniele Mestriner
Naval and Electrical
Engineering Department,
University of Genoa
Italy

Editorial Office
MDPI
St. Alban-Anlage 66
4052 Basel, Switzerland

This is a reprint of articles from the Special Issue published online in the open access journal *Applied Sciences* (ISSN 2076-3417) (available at: https://www.mdpi.com/journal/applsci/special_issues/Lightning_Modeling_Effects_Electric_Infrastructures).

For citation purposes, cite each article independently as indicated on the article page online and as indicated below:

LastName, A.A.; LastName, B.B.; LastName, C.C. Article Title. *Journal Name* **Year**, *Volume Number*, Page Range.

ISBN 978-3-0365-2626-3 (Hbk)
ISBN 978-3-0365-2627-0 (PDF)

About the Editors

Massimo Brignone was born in Finale Ligure, Italy, in 1979. He graduated cum laude in Mathematics and received his PhD in Mathematics and Applications in 2003 and in 2006 at the University of Genoa, respectively, where he is currently researcher and teacher. Massimo Brignone is co-author of more than 100 scientific contributions, published in international journals or presented at international conferences, and he has taken part in the Technical Committee of EEEIC 2019. His research interests include lightning modelling and its effects on electric infrastructures, as well as direct and indirect electromagnetic problems and microgrid energy management systems.

Daniele Mestriner was born in Genoa, Italy, on October 30, 1992. He received his M.S degree in Electrical Engineering from the University of Genoa in 2016. He is currently a Ph.D. graduating student of the Department of Electrical, Electronic and Telecommunication Engineering and Naval Architecture (DITEN) of the University of Genoa. In October 2019 he won a Post-Doc position in the framework of a collaboration between the University of Genoa, the Liguria region and Toshiba Italia. He has taken part in the Technical Committee of four international conferences: ASET 2019, CIEM 2019, EEEIC 2019 and ICOASE 2019. His main interests include lightning protection and modelling, as well as power system modelling and control. He is the author or co-author of many scientific papers published in reviewed journals or presented in international conferences.

Preface to "Lightning Modeling and Its Effects on Electric Infrastructures"

The effect of lightning on electrical infrastructures is one of the most critical issues in modern engineering research. Suitable instruments and access to up-to-date studies that are able to reflect the reality of these effects are essential. Studies relating to lightning events cover a wide variety of physics phenomena, from the development of the lightning channel, to the protection of wind turbine blades or overhead lines insulators. Presented in this chapter are works from several authors covering a range of topics, all aiming to provide better solutions to the engineering problems faced by areas of the world that are prone to lightning events.

Massimo Brignone, Daniele Mestriner
Editors

 *applied sciences*

Editorial

Lightning Modeling and Its Effects on Electric Infrastructures

Massimo Brignone and Daniele Mestriner *

Electrical, Electronics and Telecommunication Engineering and Naval Architecture Department, University of Genoa, Via All'opera Pia 11a, 16145 Genoa, Italy; Massimo.brignone@unige.it
* Correspondence: daniele.mestriner@edu.unige.it

 check for
updates

Citation: Brignone, M.; Mestriner, D. Lightning Modeling and Its Effects on Electric Infrastructures. *Appl. Sci.* **2021**, *11*, 11444. https://doi.org/10.3390/app112311444

Received: 30 November 2021
Accepted: 1 December 2021
Published: 2 December 2021

Publisher's Note: MDPI stays neutral with regard to jurisdictional claims in published maps and institutional affiliations.

1. Introduction

Infrastructure security and people's safety are the first objectives when it comes to dealing with high voltages or high currents issues. In this framework, lightning studies play a crucial role because of the dangerous consequences of this kind of phenomenon. It is well known that the normal operation of transmission and distribution systems is greatly affected by lightning, which is one of the major causes of power interruptions: lightning causes flashovers in overhead transmission and distribution lines, resulting in overvoltages on line conductors that are due either to direct strikes or to nearby, indirect strikes.

The contributions to this Special Issue mainly focused on modeling lightning activity, investigating physical causes, and discussing and testing mathematical models for the electromagnetic fields associated with lighting phenomena. In this framework, two main topics have been presented by the authors: (1) the interaction of lightning phenomena with electrical infrastructures such as wind turbines [1] and overhead lines [2–4]; and (2) the computation of lightning electromagnetic fields in the case of particular configuration, as the one presented in [5], considering a negatively charged artificial thunderstorm, in [6], considering a complex terrain with arbitrary topography, and in [7], where the ground is simplified and considered a Perfect Electric Conductor.

2. Interaction of Lightning Phenomena with Electrical Infrastructures

Wind turbines are one of most commonly damaged electrical infrastructures. The probability of being damaged increases with their height, and despite the existing lightning protection systems available for wind turbine blades, there are still many cases reported wherein damage is caused by lightning strikes. In this framework, wind turbine blades represent the most critical element of the structure, and the work proposed in [1] shows an innovative approach based on a hybrid down conductor system, which shows excellent results compared to the traditional one.

On the other hand, when we deal with lightning effects on electrical systems, researchers usually refer to overhead transmission and distribution lines. The literature has developed different numerical codes for evaluating such effects [8,9], but some open questions and unsolved doubts can still be found, especially when we deal with indirect lightning strikes. Within this category, the evaluation of the corona effect on overhead lines and a correct description of soil characteristics represent a crucial issue.

First of all, the corona effect on overhead lines requires a complex description of the relationship between the total charge and the applied voltage due to the presence of minor loops, as proposed and discussed in [4]. Secondly, in order to evaluate how the corona effect changes the number of flashovers in a distribution system, a detailed lightning performance analysis is required, as proposed by the authors of [2].

Secondly, consideration of a detailed representation of the soil and of the grounding system is extremely important as it could enhance the induced voltage. In order to solve this issue, in principle, a Full-Maxwell simulation is required, leading to high computational costs. The authors of [3] addressed this problem by introducing an equivalent circuit which takes into account all the details of the grounding grid and computing the enhancement of lightning-induced voltage compared to the traditional case [10].

3. Electromagnetic Fields Computation and Measurement

The analysis of electromagnetic fields in different configurations in terms of lightning strike and surrounding areas is crucial in order to have a comprehensive view of the phenomena.

In this framework, the analysis of upward streamer is usually neglected in the literature since it does not represent the main part of the flash. However, dealing with it is crucial in order for a monitoring system to be protected. The authors of [5] focused their efforts on replicating a physical simulation representing an upward streamer discharge and testing the spectrum of possible electromagnetic field effects on monitoring systems.

On the other hand, consideration of the principal part of lightning flashes, i.e., the return stroke, is crucial to evaluating the effect on electrical infrastructures. The research has recently divided its efforts in two main categories: (1) reducing the computational effort and (2) considering more detailed geometries for the surrounding area.

In order to reduce the computational effort, the authors of [7] provided a new method which requires a summation of analytical formulae and a simple integral operation, achieving results comparable to the one proposed in [11] and assuming a Perfect Electric Conductor ground.

The complexity of the surrounding area is taken into account by [6] thanks to an innovative open accelerator (OpenACC)-aided graphics processing unit based on the FDTD method and applied to 3D systems, which also helps in the reduction in the computational effort with respect to traditional methods based on CPU-based models.

Author Contributions: Conceptualization M.B. and D.M.; methodology, M.B. and D.M.; formal analysis, M.B. and D.M.; investigation, M.B. and D.M.; resources, M.B. and D.M.; data curation, M.B. and D.M.; writing—original draft preparation, D.M.; writing—review and editing, D.M.; supervision, M.B. All authors have read and agreed to the published version of the manuscript.

Funding: This research received no external funding.

Conflicts of Interest: The authors declare no conflict of interest.

References

1. Mucsi, V.; Ayub, A.S.; Muhammad-Sukki, F.; Zulkipli, M.; Muhtazaruddin, M.N.; Saudi, A.S.M.; Ardila-Rey, J.A. Lightning Protection Methods for Wind Turbine Blades: An Alternative Approach. *Appl. Sci.* **2020**, *10*, 2130. [CrossRef]
2. Mestriner, D.; Brignone, M. Corona Effect Influence on the Lightning Performance of Overhead Distribution Lines. *Appl. Sci.* **2020**, *10*, 4902. [CrossRef]
3. Mestriner, D.; de Moura, R.R.; Procopio, R.; Schroeder, M.D.O. Impact of Grounding Modeling on Lightning-Induced Voltages Evaluation in Distribution Lines. *Appl. Sci.* **2021**, *11*, 2931. [CrossRef]
4. Zhang, X.; Huang, K. Lightning Surge Analysis for Overhead Lines Considering Corona Effect. *Appl. Sci.* **2021**, *11*, 8942. [CrossRef]
5. Lysov, N.; Temnikov, A.; Chernensky, L.; Orlov, A.; Belova, O.; Kivshar, T.; Kovalev, D.; Voevodin, V. Physical Simulation of the Spectrum of Possible Electromagnetic Effects of Upward Streamer Discharges on Model Elements of Transmission Line Monitoring Systems Using Artificial Thunderstorm Cell. *Appl. Sci.* **2021**, *11*, 8723. [CrossRef]
6. Mohammadi, S.; Karami, H.; Azadifar, M.; Rachidi, F. On the Efficiency of OpenACC-aided GPU-Based FDTD Approach: Application to Lightning Electromagnetic Fields. *Appl. Sci.* **2020**, *10*, 2359. [CrossRef]
7. Liu, X.; Ge, T. An Efficient Method for Calculating the Lightning Electromagnetic Field Over Perfectly Conducting Ground. *Appl. Sci.* **2020**, *10*, 4263. [CrossRef]
8. Brignone, M.; Delfino, F.; Procopio, R.; Rossi, M.; Rachidi, F. Evaluation of power system lightning performance—Part II: Application to an overhead distribution network. *IEEE Trans. Electromagn. Compat.* **2016**, *59*, 146–153. [CrossRef]
9. Nucci, C.A. The lightning induced over-voltage (LIOV) code. In Proceedings of the 2000 IEEE Power Engineering Society Winter Meeting, Conference Proceedings (Cat. No. 00CH37077). Singapore, 23–27 January 2000; Volume 4, pp. 2417–2418.

10. Brignone, M.; Delfino, F.; Procopio, R.; Rossi, M.; Rachidi, F. Evaluation of power system lightning performance, part i: Model and numerical solution using the pscad-emtdc platform. *IEEE Trans. Electromagn. Compat.* **2016**, *59*, 137–145. [CrossRef]
11. Mestriner, D.; Brignone, M.; Procopio, R.; Rossi, M.; Delfino, F.; Rachidi, F.; Rubinstein, M. Analytical Expressions for Lightning Electromagnetic Fields With Arbitrary Channel-Base Current. Part II: Validation and Computational Performance. *IEEE Trans. Electromagn. Compat.* **2020**, *63*, 534–541. [CrossRef]

Article

Lightning Surge Analysis for Overhead Lines Considering Corona Effect

Xiaoqing Zhang * and Kejie Huang

School of Electrical Engineering, Beijing Jiaotong University, Beijing 100044, China; 12121491@bjtu.edu.cn
* Correspondence: zxqing002@163.com

Abstract: Corona discharge characteristics are measured in a corona cage. The difference is found between the q–u curves under double exponential and damped oscillation surges. The behavior of the minor loops is revealed for the q–u curves under positive and negative damped oscillation surges. An extended improvement is made on the traditional approach for modeling of the q–u curves under damped oscillation surges. The extended approach has the capability of describing the complicated trajectory feature of the minor loops. On the basis of the extended approach, an efficient method is proposed for performing lightning surge analysis of overhead lines considering the corona effect. In the proposed method, an overhead line with corona is divided into a certain number of line segments. Each segment is converted into a circuit unit consisting of a non-linear branch and a linear circuit. With these circuit units connected in sequence, a complete equivalent circuit is constructed for the overhead line with corona. The transient responses can be obtained from the solution to the equivalent circuit. Then, the calculated results are compared with the field test results on a test overhead line.

Keywords: corona; lightning surge; overhead line; transient calculation

check for **updates**

Citation: Zhang, X.; Huang, K.
Lightning Surge Analysis for
Overhead Lines Considering Corona
Effect. *Appl. Sci.* **2021**, *11*, 8942.
https://doi.org/10.3390/
app11198942

Academic Editors: Massimo Brignone
and Daniele Mestriner

Received: 31 August 2021
Accepted: 22 September 2021
Published: 25 September 2021

1. Introduction

Analysis of the propagation behavior of lightning surges on overhead lines is essential for lightning overvoltage protection and insulation coordination of electric power apparatus. It is well known that lightning surges undergo distortion and attenuation when they propagate along overhead lines. Surge corona plays a significant part in the distortion and attenuation phenomena. The design of lightning overvoltage protection depends strongly on accurate knowledge of the amplitude and wavefront steepness of lightning surges [1–6]. In the calculation of lightning overvoltages, surge corona is taken into account by its charge-voltage characteristic, namely the q–u curve. Measurements of the q–u curves have been taken in the corona cages and on the actual overhead lines [7–10]. Nevertheless, the measured data were obtained mainly under double exponential surges. The distortion and attenuation were always calculated according to the q–u curves under double exponential surges, no matter what the actual surge waveshapes are [3,11–13]. As a matter of fact, the great majority of lightning surges intruding into substations are damped oscillation surges due to the refraction and reflection of surge waves [14,15]. There is a practical need for taking into account the corona effect on the distortion and attenuation of damped oscillation surges. Some work suited to this need was reported in literature [16–19]; however, the systematic research results were still not reported in the previous work. In view of this situation, an attempt is made in this paper to comprehensively investigate the corona effect on the distortion and attenuation of damped oscillation surges. An experimental measurement is made in a corona cage. The basic difference is found between the q–u curves under double exponential and damped oscillation surges. The trajectory feature of the minor loops is also observed under positive and negative damped oscillation surges. Owing to the fact that the traditional approach is only suitable to modeling of the monotonic q–u curves under double exponential surges [20], an extended approach is presented for modeling of the hysteresis-like q–u curves under damped oscillation surges. With the extended approach implemented into the transient analysis, an efficient method is proposed to calculate the lightning transients on overhead lines. The proposed method can effectively predict the distortion and attenuation of damped oscillation surges on overhead lines with corona. The calculated results are compared with the field test results to check the validity of the proposed method. Then, we further discuss the calculated results in the presence and absence of the minor loops to inquire into the influence of the minor loops on the distortion and attenuation of damped oscillation surges under positive and negative polarities.

2. Experimental Investigation on Corona Characteristics

An experimental arrangement was built in a high voltage laboratory for measuring the corona q–u curves, as shown in Figure 1. Its schematic diagram is illustrated in Figure 2. Double exponential and damped oscillation surge voltages can be generated by putting the wavefront resistor R_W and inductor L_W into operation in the impulse generator IG, respectively. The corona cage consists of an inner electrode and three sections of outer electrodes. These electrodes are coaxially assembled with the longer outer electrode E_L (1 m) in the middle and the two shorter outer electrodes E_S (0.52 m) at both sides to shield the end effect of the electric field. When a surge voltage is applied to the inner electrode E_i, corona discharge is produced in the corona cage. The charge signal q and voltage signal u are taken from the integral capacitor C_e and voltage divider VD, respectively. The two signals are recorded by a digital oscilloscope DS, so that the q–u curves can be obtained from the data processing for the signals q and u. For the sake of comparison, the wavefront time and amplitude are held to be approximately equal for double exponential and damped oscillation surges. Figure 3 shows a group of measured q–u curves under negative and positive polarities. It can be noticed that the q–u curves under damped oscillation surges roughly coincide with those under double exponential surges until their respective charges reach the maximum values. The difference appears between the curve parts subsequent to the maximum charge points. In these parts, the q–u curves under double exponential surges descend monotonically; however, those under damped oscillation surges follow a hysteresis-like trajectory. The possible forming mechanism of the hysteresis-like trajectory might be attributed to the occurrence of the opposite polar corona [21] and is illustrated by Figure 4. With the voltage u decreasing to the first wave trough, the slope of the section FG becomes larger than the geometrical capacitance C_0, which causes the section FG to deviate from the section EF. In fact, as the voltage u decreases to a certain extent, the electric field near the surface of the inner electrode could be reversed. Once the reversed field strength exceeds a critical value, the opposite polar corona discharge may occur, which produces the opposite polar space charge near the inner electrode. Thence a reduction in the net space charge leads the total charge q on the section FG to decrease more rapidly than that on the section EF. A similar interpretation can be reached for the section HP as the voltage u decreases to the second wave trough. For this reason, the minor loops is formed after the first oscillation cycle. The minor loops enclosed by the hysteresis-like trajectory are significantly larger under negative polarity than under positive polarity. The area of the minor loops represents the energy loss and can distort and attenuate damped oscillation surges in the subsequent oscillation cycles.

Figure 1. Experimental arrangement in high voltage laboratory.

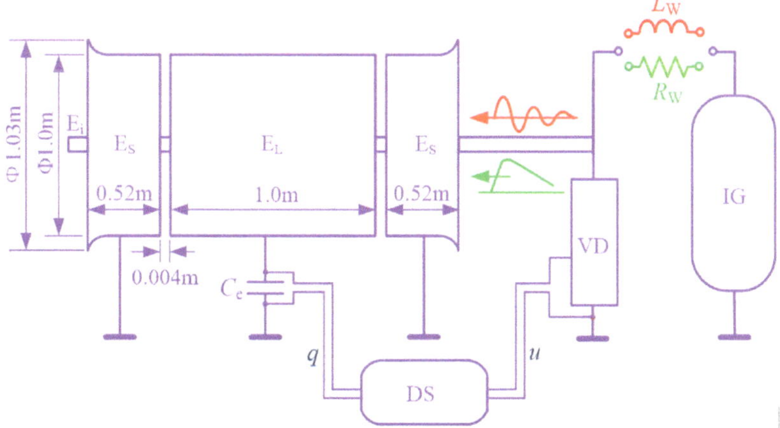

Figure 2. Diagram of experimental system (E_i—inner electrode; E_L—longer outer electrode; E_S—shorter outer electrode; DS—digital oscilloscope; VD—voltage divider; IG—impulse generator)

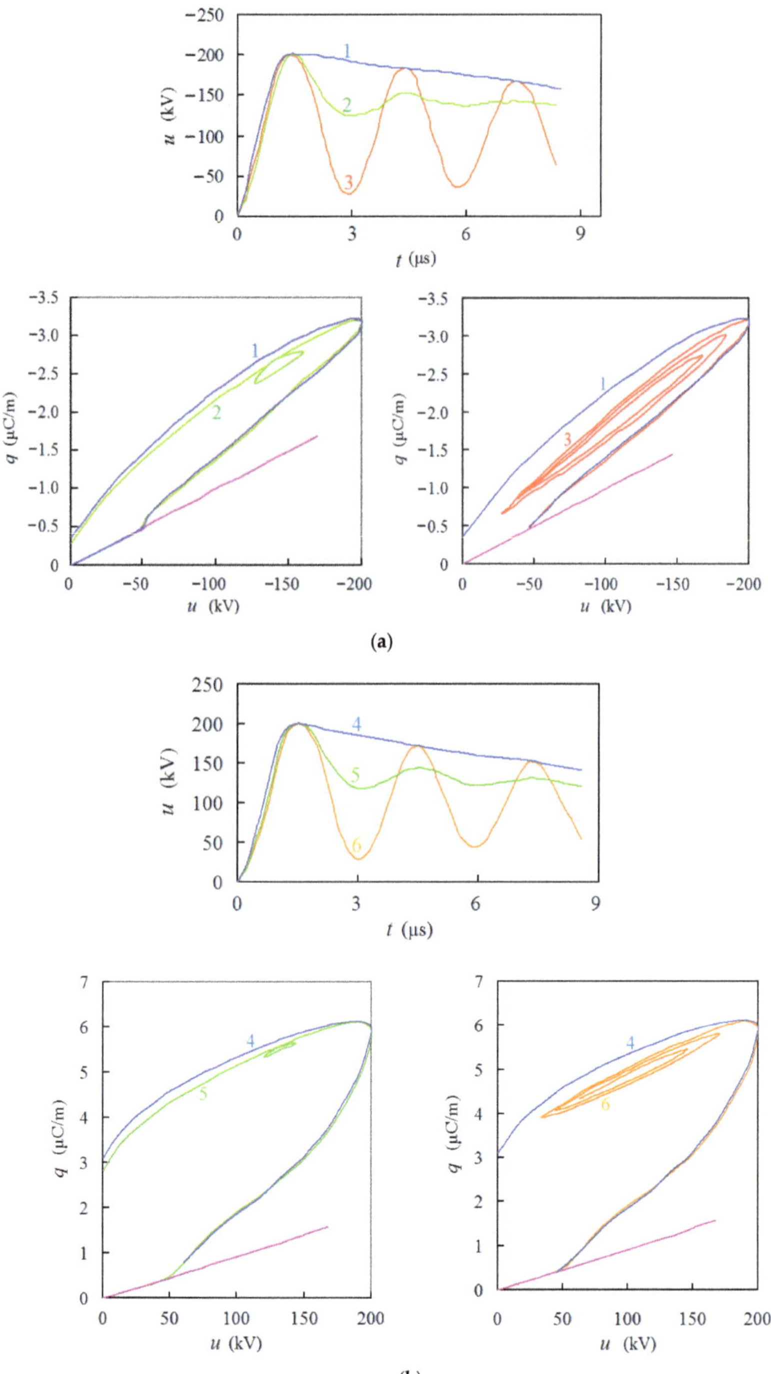

Figure 3. Measured q–u curves. (**a**) Under negative polarity, (**b**) under positive polarity.

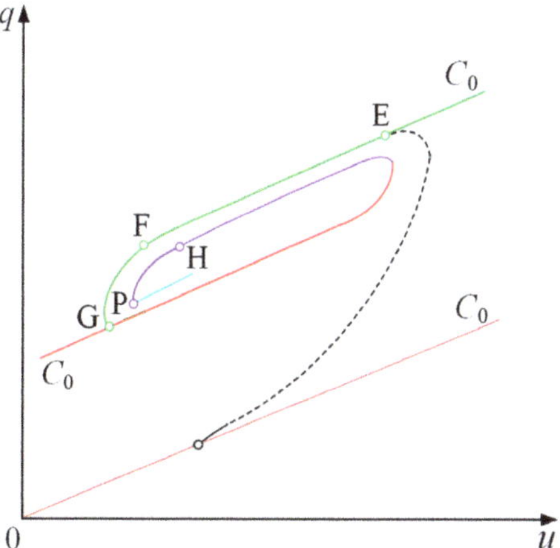

Figure 4. Formulation of minor loops.

3. Modeling of Corona q–u Curves

The traditional approach is only suitable for modeling of the q–u curves under double exponential surges [20]. It introduced the corona charge q_C by subtracting the induced charge C_0u from the total charge q [20,22,23], as illustrated in Figure 5. The corona current is described by:

$$i_C = \begin{cases} 0 & q_C > (C_2 - C_0)(u - U_2) \\ f(u, q_C) & (C_1 - C_0)(u - U_1) < q_C < (C_2 - C_0)(u - U_2) \\ h(u, q_C) & 0 < q_C < (C_1 - C_0)(u - U_1) \end{cases} \tag{1}$$

where,

$$\begin{aligned} f(u, q_C) &= \alpha[(C_2 - C_0)(u - U_2) - q_C] \\ h(u, q_C) &= f(u, q_C) + \beta[(C_1 - C_0)(u - U_1) - q_C] \end{aligned} \tag{2}$$

where C_1 and C_2 are the dynamic capacitances ($C_2 > C_1 > C_0$); U_1 and U_2 are the critical voltages ($U_1 > U_2$); α and β are the coefficients ($\alpha > 0$, $\beta > 0$), as shown in Figure 6.

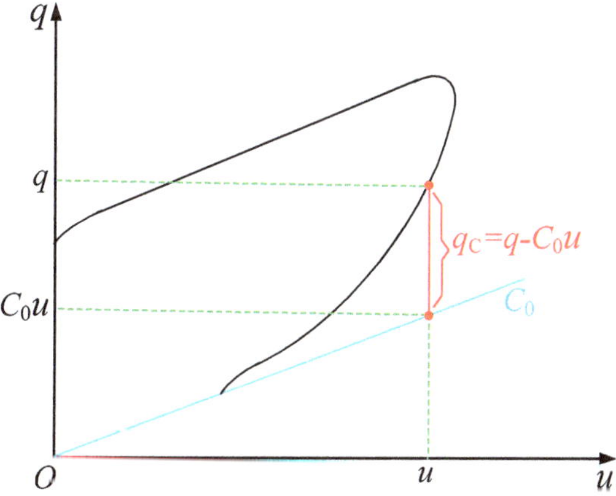

Figure 5. Diagram of corona charge.

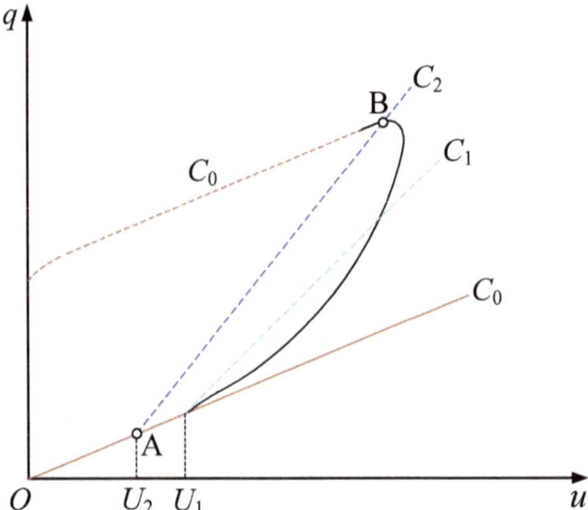

Figure 6. Model parameters.

In view of the trajectory complexity of the q–u curves under damped oscillation surges, an extended improvement is made on the traditional approach [20,23]. For such a q–u curve, it is divided into different curve segments, as shown in Figure 7. The first curve segment AB is still described by Equation (1). For modeling of the second curve segment BB′, the origin of coordinate system is translated to point B, where $u' = u - U_B$ and $q' = q - Q_B$. In this way, the curve segment BB′ is described by:

$$i_C = \begin{cases} 0 & q'_C > (C'_2 - C_0)(u' - U'_2) \\ f(u', q'_C) & (C'_1 - C_0)(u' - U'_1) < q'_C < (C'_2 - C_0)(u' - U'_2) \\ h(u', q'_C) & 0 < q'_C < (C'_1 - C_0)(u' - U'_1) \end{cases} \tag{3}$$

where,

$$\begin{aligned} f(u', q'_C) &= \alpha'[(C'_2 - C_0)(u' - U_2) - q'_C] \\ h(u', q'_C) &= f(u', q'_C) + \beta'[(C'_1 - C_0)(u' - U'_1) - q'_C] \end{aligned} \tag{4}$$

where $C'_2 > C'_1 > C_0$, $U'_1 > U'_2$, $\alpha' > 0$ and $\beta' > 0$. In a similar manner, the third curve segment B′B″, as shown in Figure 8, is described by:

$$i_C = \begin{cases} 0 & q''_C > (C''_2 - C_0)(u'' - U''_2) \\ f(u'', q''_C) & (C''_1 - C_0)(u'' - U''_1) < q''_C < (C''_2 - C_0)(u'' - U''_2) \\ h(u'', q''_C) & 0 < q''_C < (C''_1 - C_0)(u'' - U''_1) \end{cases} \tag{5}$$

where,

$$\begin{aligned} f(u'', q''_C) &= \alpha''[(C''_2 - C_0)(u'' - U''_2) - q''_C] \\ h(u'', q''_C) &= f(u'', q''_C) + \beta''[(C''_1 - C_0)(u'' - U''_1) - q''_C] \end{aligned} \tag{6}$$

where $C''_2 > C''_1 > C_0$, $U''_1 > U''_2$, $\alpha'' > 0$ and $\beta'' > 0$.

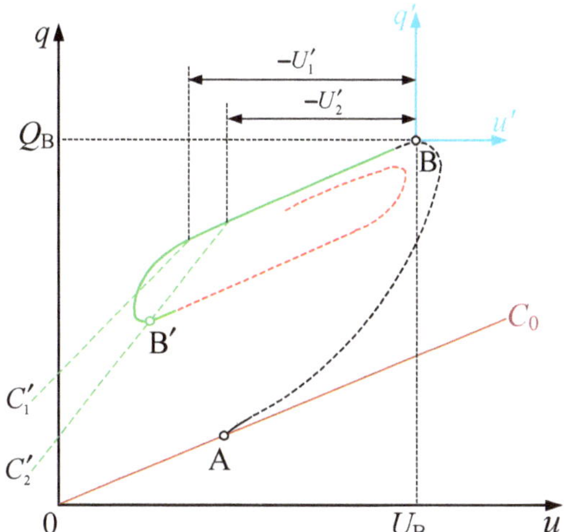

Figure 7. Modeling of the second curve segment.

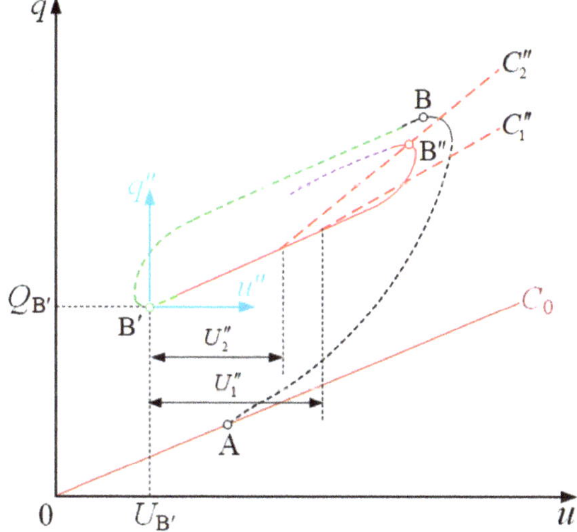

Figure 8. Modeling of the third curve segment.

Description of the subsequent curve can be made by analogy with that of either the curve segment BB′ or the curve segment B′B″. In Equations (1)–(6), the parametric values of $\alpha \sim \alpha''$, $\beta \sim \beta''$, $C_1 \sim C_2''$, $U_1 \sim U_1''$ and $U_2 \sim U_2''$ can be determined by fitting the given q–u curve. The integral of the corona current i_C from $t - \Delta t$ to t gives the corona charge

$$q_C(t) = \frac{\Delta t}{2} i_C(t) + q_{CP}(t - \Delta t) \tag{7}$$

where Δt is the time step and $q_{CP}(t - \Delta t)$ is corona charge at the preceding time step:

$$q_{CP}(t - \Delta t) = \frac{\Delta t}{2} i_C(t - \Delta t) + q_C(t - \Delta t) \tag{8}$$

According to Figure 3, the total charge q on the q–u curve is evaluated by:

$$q(t) = q_C(t) + C_0 u(t) \tag{9}$$

4. Transient Calculation Considering Corona Effect

Figure 9a shows an overhead line with corona. It is subdivided into M line segments (see Figure 9b). On each line segment, the corona sheath is approximately considered to be uniform and replaced as a lumped non-linear branch carrying a corona current [24,25], as shown in Figure 9c. After separating the corona sheath from each line segment, the remainder is free of corona and has the linear circuit parameters per unit length, i.e., R_1, L_1 and C_1. Its equivalent circuit is depicted in Figure 10 [26], where $\Delta R = \Delta l R_1$, $\Delta L = \Delta l L_1$, $\Delta C = \Delta l C_1$, $Z = (\Delta L / \Delta C)^{1/2}$ and $\tau = \Delta l / v$ (v is the wave velocity). The expressions of the historical current sources $I_{j-1}(t - \tau)$ and $I_j(t - \tau)$ were also given in [26]. Considering Figures 9c and 10, the overhead line with corona is converted into a complete equivalent circuit, as shown in in Figure 11. This is a cascade circuit containing M non-linear branches carrying corona current. Topologically each unit including the non-linear branch is disconnected from other. The jth ($j = 1, 2, \cdots, M$) circuit unit in Figure 11 can be represented as a one–port circuit, as shown in Figure 12. With the non-linear branch removed from the port (see Figure 13), the corresponding open-circuit voltage u_{OCj} and input impedance Z_{thj} is found by using the Thevenin's equivalent technique. The non-linear branch can be solved according to the Thevenin's equivalent circuit shown in Figure 14:

$$u_{OCj} - Z_{thj}i_{cj} = u_{Cj}(j = 1, 2, \cdots, M) \tag{10}$$

where the corona current i_{Cj} is described by Equations (1), (3) and (5). The fundamental studies made on an equivalent circuit model may refer to [27–29]. For the first curve section on the q–u curve, as described by Equation (1), if the corona charge satisfies $(C_1 - C_0)(u_j - U_1) < q_{cj} < (C_2 - C_0)(u_j - U_2)$, the corona current and branch voltage are given by:

$$i_{Cj} = \frac{\beta}{1+\beta\zeta}[(C_2 - C_0)(u_{Cj} - U_2) - q_{CP}(t - \Delta t)]$$
$$u_{Cj} = \frac{1}{1+\Gamma}[u_{oc} + \Gamma U_2 - \frac{\Gamma}{C_2 - C_0}q_{CP}(t - \Delta t)] \tag{11}$$

where $\zeta = \Delta t / 2$ and Γ is,

$$\Gamma = \frac{\beta Z_{thj}(C_2 - C_0)}{1 + \beta\zeta} \tag{12}$$

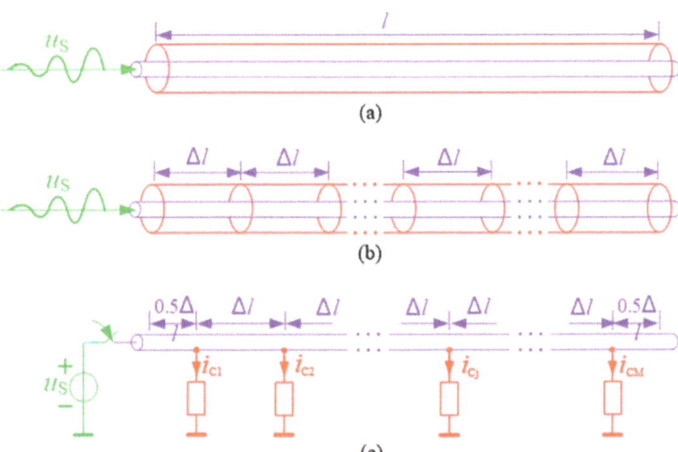

Figure 9. Segmentation of an overhead line with corona. (**a**) An overhead line with corona, (**b**) line segment with corona, (**c**) non-linear branch carrying corona current.

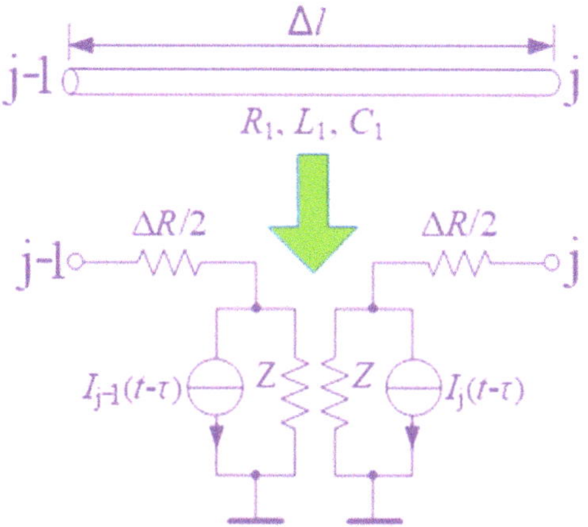

Figure 10. Equivalent circuit of a line segment free of corona.

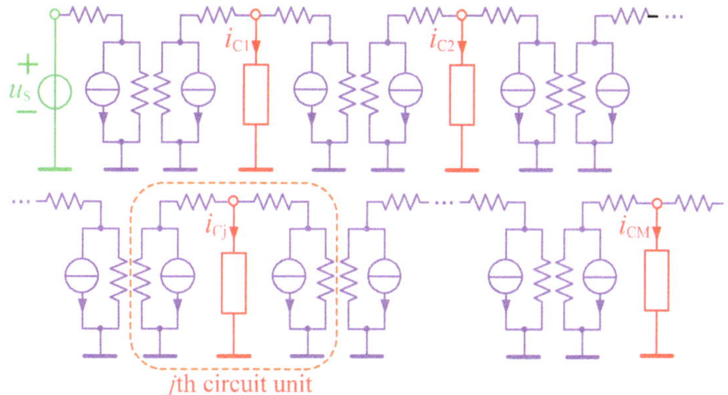

Figure 11. Complete equivalent circuit of an overhead line with corona.

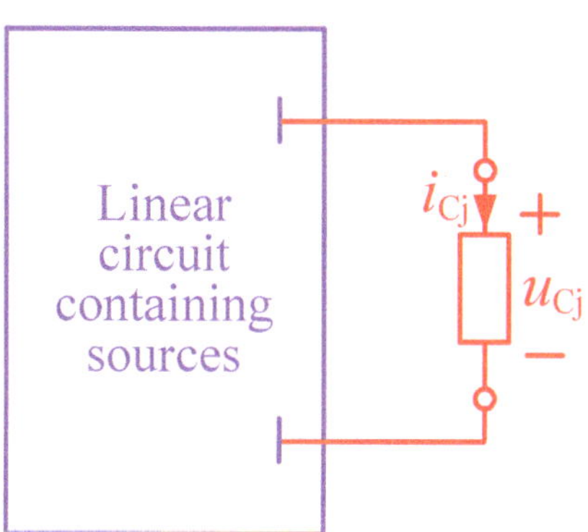

Figure 12. One-port circuit containing nonlinear branch.

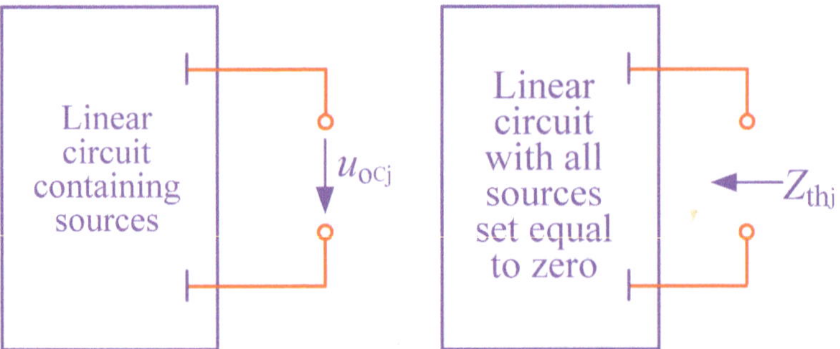

Figure 13. Thevenin's equivalent procedure.

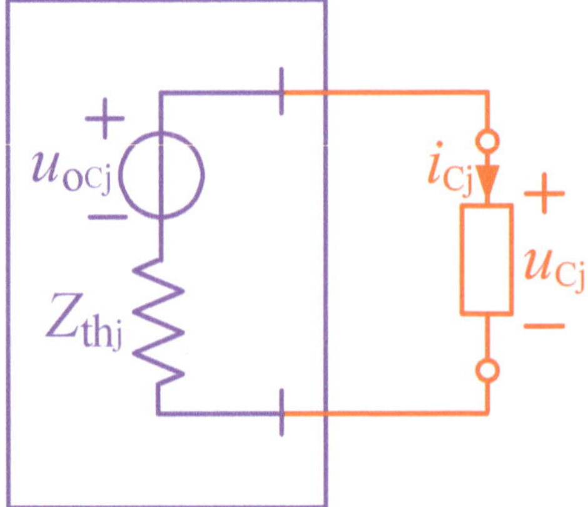

Figure 14. Solution to non-linear branch.

If the corona charge satisfies $0 < q_{cj} < (C_1 - C_0)(u_j - U_1)$, the corona current and branch voltage are given by:

$$i_{Cj} = \frac{1}{\Lambda_1}[\alpha(C_1 - C_0)(u_{Cj} - U_1) + \beta(C_2 - C_0)(u_{Cj} - U_2) - (\alpha + \beta)q_{CP}(t - \Delta t)]$$
$$u_{Cj} = \frac{1}{\Lambda_1 + \Lambda_2 + \Lambda_3}[\Lambda_1 u_{oCj} + \Lambda_2 U_1 + \Lambda_3 U_2 + \Lambda_4 q_{CP}(t - \Delta t)] \tag{13}$$

where,

$$\Lambda_1 = 1 + \xi(\alpha + \beta)$$
$$\Lambda_2 = Z_{thj}\alpha(C_1 - C_0)$$
$$\Lambda_3 = Z_{thj}\beta(C_2 - C_0)$$
$$\Lambda_4 = Z_{thj}(\alpha + \beta) \tag{14}$$

For the second, third and subsequent curve segments on the q–u curve, the corresponding solutions of the corona current and branch voltage can be obtained in a manner similar to Equations (12) and (14). After determining i_{Cj} and u_{Cj} ($j = 1, 2, \cdots, M$), the nonlinear branch is replaced by a voltage source equal to u_{Cj}. Figure 12 is converted into a purely linear circuit, as shown in Figure 15, and thereby the transient responses in the remaining linear part can be calculated by performing the nodal voltage analysis. The calculation procedure has been stated in detail in [26,30,31] and is not repeated here due to the limitation of space.

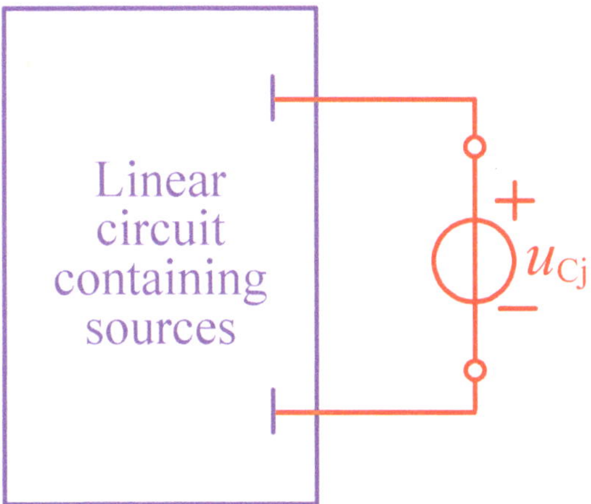

Figure 15. Solution to remaining linear circuit.

5. Calculated Results and Discussions

A test overhead line is considered here [17]. Its line conductor is a single-core copper wire with a cross-section of 16 mm^2. The average line height is 10 m above the ground and total length l is 3084 m. The geometrical capacitance C_0 is 6.41 pF/m and the soil resistivity is about 50 $\Omega\cdot$m. The corona onset voltages under positive and negative polarities are taken as 70 kV and 80 kV, respectively. The damped oscillation surge voltages with different amplitudes and polarities are applied to the sending end of the line. Two capacitive voltage dividers are installed at the sending end and at 1674 m from the sending end, respectively, to measure the voltage waveshapes. The parameter values for describing the q–u curves are given in Table 1, where the signs (+) and ($-$) denote positive and negative polarities. Using the method proposed above, the voltage waveshapes are calculated at a distance of 1674 m from the sending end, as shown in Figures 16 and 17. The calculated voltage values at the first two wave crests and first wave trough are listed in Tables 2 and 3. In Figures 16 and 17 and Tables 2 and 3, the field test results [17] and those calculated in the presence and absence of the minor loops are given together for comparison, which shows the calculated results can agree well with the field test results. This confirms the validity of the proposed method.

Table 1. Parameter values of modeling q–u curves.

	α	α'	α''	β	β'	β''
	(MHz)				(kHz)	
(+)	0.4	4.5	1.5	110.0	1.1	110.0
($-$)	8.0	4.5	15.0	210.0	1.1	110.0
	C_1	C_2	C_1'	C_1''	C_2'	C_2''
	(pF/m)					
(+)	13.5	20.0	6.2	8.0	6.3	14.0
($-$)	7.5	8.0	6.5	6.5	6.5	7.0
	U_1	U_2	U_1'	U_1''	U_2'	U''
	(kV)					
(+)	95.0	90.0	-5.0	45.0	-4.0	35.0
($-$)	90.0	85.0	-5.0	55.0	-4.0	50.0

Table 2. Voltage values under positive polarity (kV).

Applied voltage amplitude		156	178	199	226	254	275	319	328
The first wave crest	Calculated	87.4	114.2	116.5	117.8	120.6	127.3	136.7	142.8
	Field test	84.5	112.5	114.7	120.3	126.5	131.7	140.8	144.7
The first wave trough	Calculated (in the presence of minor loops)	55.2	60.3	70.2	82.2	84.0	92.8	109.6	110.2
	Calculated (in the absence of minor loops)	52.6	55.2	68.1	78.9	82.2	91.3	104.7	106.8
	Field test	55.1	56.8	67.9	82.0	86.4	94.4	110.6	117.1
The second wave crest	Calculated (in the presence of minor loops)	104.1	117.1	134.9	153.6	158.0	170.0	207.9	237.8
	Calculated (in the absence of minor loops)	106.8	118.2	136.3	157.4	165.3	171.7	215.5	241.2
	Field test	105.2	114.4	135.8	155.7	157.1	170.4	212.3	227.1

Table 3. Voltage values under negative polarity (kV).

Applied voltage amplitude		175	188	195	226	258	292	340	358
The first wave crest	Calculated	128.3	149.2	151.1	186.1	200.3	224.6	263.4	271.9
	Field test	125.7	147.5	149.7	181.8	197.2	222.5	261.6	269.5
The first wave trough	Calculated in the presence of minor loops)	58.7	64.9	67.3	86.7	99.1	104.5	117.5	134.8
	Calculated (in the absence of minor loops)	56.5	62.3	65.8	79.4	90.7	98.6	105.8	119.5
	Field test	59.6	68.1	70.7	89.4	104.2	106.7	116.5	133.0
The second wave crest	Calculated (in the presence of minor loops)	129.5	139.3	142.9	166.9	193.8	214.9	230.1	250.5
	Calculated (in the absence of minor loops)	134.8	143.1	145.7	170.1	198.4	225.2	241.3	261.4
	Field test	128.3	136.0	136.7	163.2	190.1	215.2	218.2	242.2

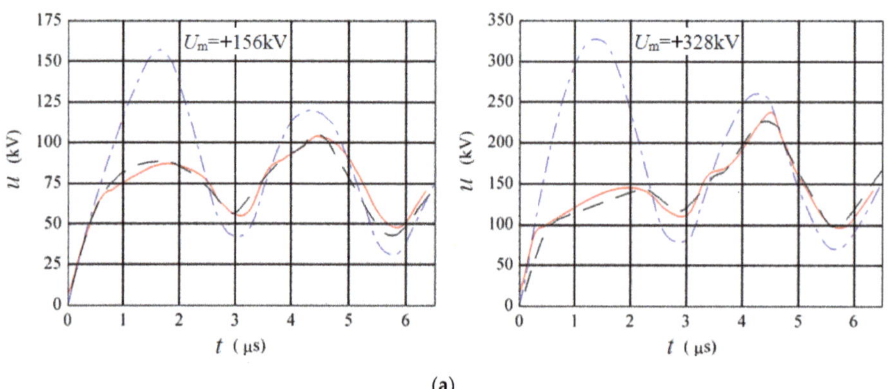

(a)

Figure 16. *Cont.*

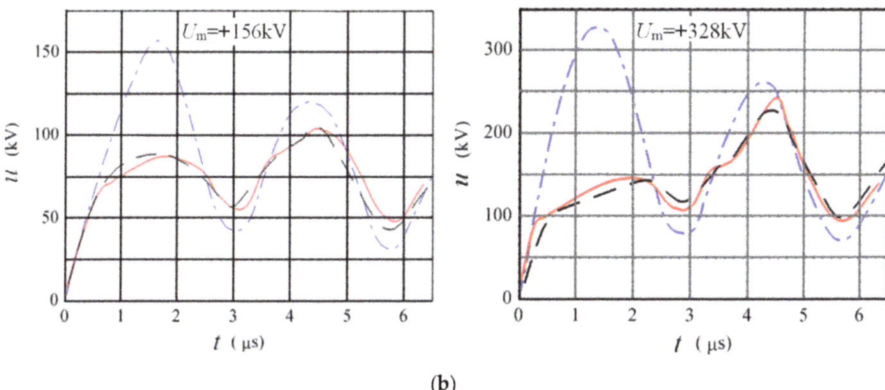

(b)

Figure 16. Calculated and field test voltage waveshapes under positive polarity (U_m is the voltage amplitude at the sending end; — · — · — sending end; — — — field test; ———— calculated). (**a**) In the presence of minor loops, (**b**) in the absence of minor loops.

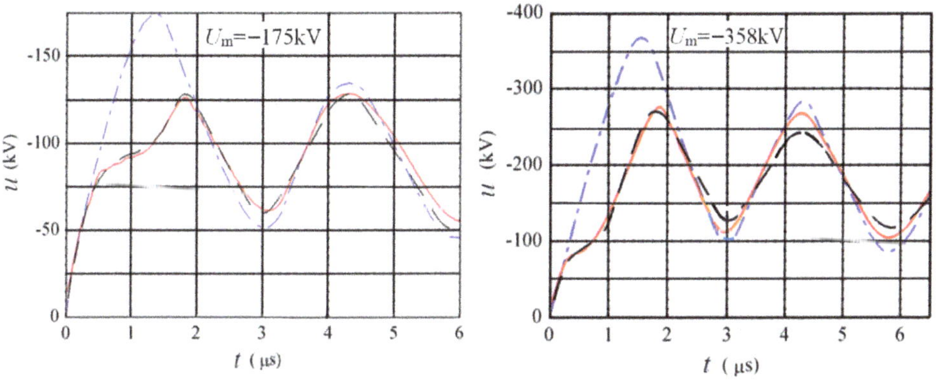

(a)

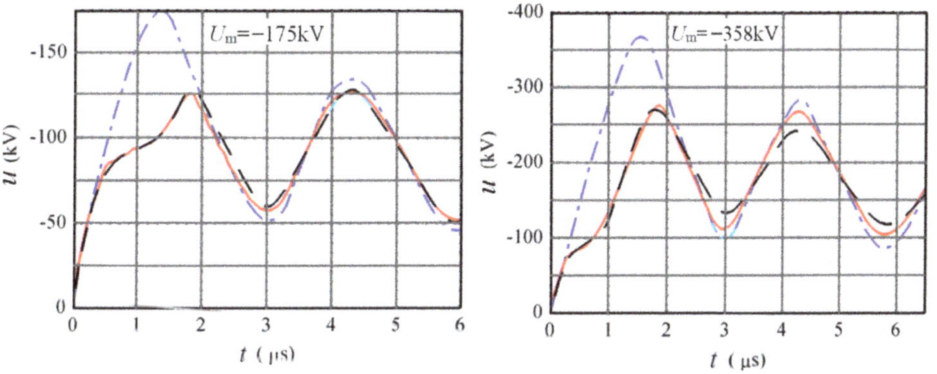

(b)

Figure 17. Calculated and field test voltage waveshapes under negative polarity (U_m is the voltage amplitude at the sending end; — · — · — sending end; — — — field test; ———— calculated). (**a**) In the presence of minor loops, (**b**) in the absence of minor loops.

As can be seen from Figures 16 and 17 and Tables 2 and 3, the distortion and attenuation of the surge voltages in the first oscillation cycle are considerably more severe under positive polarity than under negative polarity. This is due to the fact that the area of the main loop formed by the curve in the first oscillation cycle is much larger under positive polarity than under negative polarity according to the measured q–u curves shown in Figure 3. Hence the corona discharge in the first oscillation cycle can cause much higher energy loss under positive polarity than under negative polarity. The attenuation decrement of the first wave crest is 44.0~56.4% under positive polarity, whereas that is only 24.1~26.7% under negative polarity. In addition, the minor loops also have different influences on calculation of the distortion and attenuation of the surge voltages. Under positive polarity, as shown in Figure 16 and Table 2, the calculated errors (relative to the field test results) at the first wave trough and second wave crest are 1.7~6.1% and 1.1~4.7% in the presence of the minor loops and 2.8~6.7% and 1.5~4.5% in the absence of the minor loops, respectively. It is thus clear that the calculated errors in the presence and absence of the minor loops are close to each other. The reason is that the area of the minor loops is relatively small on the q–u curves under positive polarity in terms of the measured results given above, and incapable of producing considerable energy loss for the distortion and attenuation after the first oscillation cycle. However, the case under negative polarity is different from that under positive polarity. As shown in Figure 17 and Table 3, the calculated errors at the first wave trough and second crest reach 5.2~10.2% and 5.1~10.6% in the absence of the minor loops, whereas those are only 1.5~4.8% and 0.94~5.4% in the presence of the minor loops, respectively. It follows that there is an appreciable difference in the calculation precision in the presence and absence of the minor loops. Moreover, the calculated results in the presence of the minor loops more closely approximate the field test results. This can be interpreted as the area of minor loops being relatively large on the q–u curves under negative polarity in the light of the aforementioned experimental observation and causing considerable energy loss to distort and attenuate the surge voltages in the subsequent oscillation cycles. Therefore, neglect of the minor loops in the transient calculation may give rise to non-negligible error for predicting the distortion and attenuation of negative damped oscillation surges.

For a further verification of the proposed method, the calculated voltage waveshapes are also compared with those obtained from the FDTD method [32], as shown in Figure 18. On the whole, the former agrees with the latter and both are close to the field test voltage waveshapes.

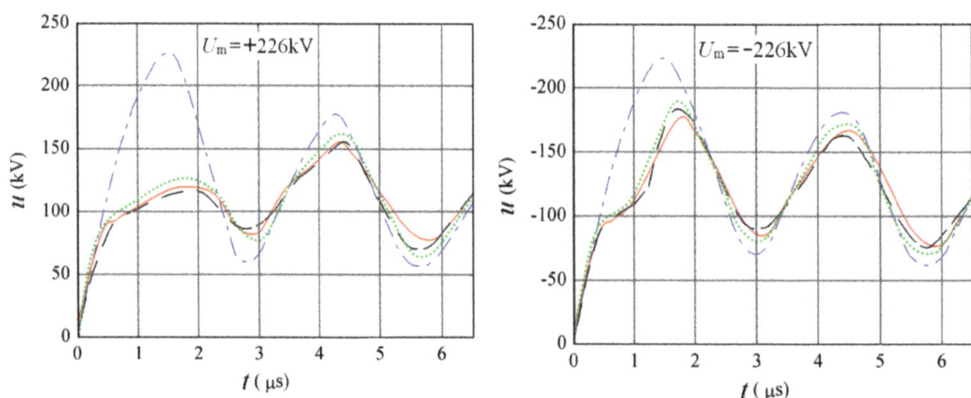

Figure 18. A comparison with FDTD method (U_m is the voltage amplitude at the sending end; — · — · — sending end; — — — field test; ———— proposed method; ············ FDTD).

6. Conclusions

The corona characteristics have been measured experimentally in a corona cage under damped oscillation and double exponential surges. The measured results shows that a basic difference between the q–u curves under the two types of surge voltage appears after their respective maximum charges. In these curve parts, the q–u curves under double exponential surges descend monotonically, whereas those under damped oscillation surges behave as the minor loops. The area of the minor loops is significantly larger under negative polarity than under positive polarity. In order to characterize the complicated trajectory feature of the minor loops, an extended approach has been presented by considerably improving the traditional approach. The extended approach is adaptable not only to double exponential surges but also to damped oscillation surges. Based on the extended approach, an efficient method has been proposed for calculating the distortion and attenuation of damped oscillation surges on the overhead lines with corona. A better agreement between the calculated and field test results confirms the validity of the proposed method. Furthermore, special emphasis is put on the fact that the influence of the minor loops on the distortion and attenuation of damped oscillation surges is obviously stronger under negative polarity than under positive polarity in the subsequent oscillation cycles. The calculated results under negative polarity indicate that the calculated errors due to neglecting the minor loops may reach 10.2~10.6% at the first wave trough and second crest, respectively. Since most lightning surges have the negative polarity, it is of practical significance to take into account the influence of the minor loops in the transient calculation. In future work, a further improvement will be made to the proposed method in the lightning transient calculation of multiphase overhead lines with bundle conductors.

Author Contributions: Conceptualization, X.Z.; methodology, X.Z. and K.H.; software, K.H.; validation, K.H.; formal analysis, X.Z.; investigation, K.H.; resources, K.H.; data curation, K.H.; writing—original draft preparation, X.Z.; writing—review and editing, X.Z.; visualization, K.H.; supervision, X.Z.; project administration, X.Z. Both authors have read and agreed to the published version of the manuscript.

Funding: This research was funded by National Natural Science Foundation of China, grant number 51777007.

Institutional Review Board Statement: No applicable.

Informed Consent Statement: No applicable.

Data Availability Statement: The data present in this study are available on request from the corresponding author.

Conflicts of Interest: The authors declare no conflict of interest.

References

1. Said, A.; Anane, Z. Corona lightning overvoltage analysis for 500kV hybrid line. *IET Gener. Transm. Distrib.* **2020**, *14*, 532–541. [CrossRef]
2. Mestriner, D.; Brignone, M. Corona effect influence on the lightning performance of overhead distribution lines. *Appl. Sci.* **2020**, *10*, 4902. [CrossRef]
3. Yutthagowitha, P.; Tran, T.H.; Baba, Y.; Ametani, A.; Rakov, V.A. PEEC simulation of lightning overvoltage surge with corona discharges on the overhead wires. *Electr. Power Syst. Res.* **2020**, *180*, 106118. [CrossRef]
4. Abouelatta, M.A.; Ward, S.A.; Sayed, A.M.; Mahmoud, K.; Lehtonen, M.; Darwish, M.M.F. Measurement and assessment of corona current density for HVDC bundle conductors by FDM integrated with full multigrid technique. *Electr. Power Syst. Res.* **2021**, *199*, 107370. [CrossRef]
5. Abouelatta, M.A.; Ward, S.A.; Sayed, A.M.; Mahmoud, K.; Lehtonen, M.; Darwish, M.M.F. Fast corona discharge assessment using FDM integrated with full multigrid method in HVDC transmission lines considering wind Impact. *IEEE Access* **2020**, *8*, 225872–225883. [CrossRef]
6. Abdel Gawad, N.M.K.; Shaalan, E.M., Darwish, M.M.F.; Basuny, M.A.M. Influence of fault locations on the pipeline induced voltages near to power transmission lines. In Proceedings of the 2019—21st International Middle East Power Systems Conference (MEPCON), Cairo, Egypt, 17–19 December 2019; pp. 761–767.

7. Yang, P.C.; Chen, S.M.; He, J.L. Lightning impulse corona characteristic of 1000-kV UHV transmission lines and its influences on lightning overvoltage analysis results. *IEEE Trans. Power Deliv.* **2013**, *28*, 2518–2525. [CrossRef]
8. Maruvad, P.S.; Menemenlis, H.; Malewaki, R. Corona characteristics of conductor bundles under impulse voltages. *IEEE Trans. Power App. Syst.* **1977**, *96*, 102–114. [CrossRef]
9. Podporkin, G.V.; Sivaev, A.D. Lightning impulse characteristics of conductors and bundles. *IEEE Trans. Power Deliv.* **1997**, *12*, 1842–1847. [CrossRef]
10. Bochkovskii, V.V. Impulse corona on single and bundled conductors. *Elektrichestvo* **1966**, *7*, 22–27.
11. Carneiro, S.; Marti, J.R. Evaluation of corona and line models in electromagnetic transients simulations. *IEEE Trans. Power Deliv.* **1991**, *6*, 334–342. [CrossRef]
12. He, J.L.; Zhang, X.; Yang, P.H.; Chen, S.M.; Zeng, R. Attenuation and deformation characteristics of lightning impulse corona traveling along bundled transmission lines. *Electr. Power Syst. Res.* **2015**, *118*, 29–36. [CrossRef]
13. Araneo, R.; Maccioni, M.; Lauria, S.; Geri, A.; Gatta, F.M.; Celozzi, S. Comparison of corona models for computing the surge propagation in multiconductor power lines. In Proceedings of the 2016 IEEE 16th International Conference on Environment and Electrical Engineering (EEEIC), Florence, Italy, 7–10 June 2016; pp. 1–6.
14. Ozawa, J.; Ohsaki, E.; Ishii, M.; Kojima, S.; Ishihara, H.; Kouno, T.; Kawamura, T. Lightning surge analysis in a multi-conductor system for substation insulation design. *IEEE Trans. Power App. Syst.* **1985**, *PAS-104*, 2244–2254. [CrossRef]
15. Takami, J.; Okabe, S.; Zaima, E. Study of lightning surge overvoltages at substations due to direct lightning strokes to phase conductors. *IEEE Trans. Power Deliv.* **2010**, *25*, 425–433. [CrossRef]
16. Rickard, D.A.; Harid, N.; Waters, R.T. Modelling of corona at a high-voltage conductor under double exponential and oscillation impulses. *IEE Proc.-Sci. Meas. Technol.* **1996**, *143*, 277–284. [CrossRef]
17. Zayients, S.L.; Kostenko, M.V.; Lyapin, A.G. An experimental investigation on the deformation of travelling wave due to impulse corona. *Bull. Polytech. Inst. Leningr.* **1958**, *195*, 342–372.
18. Hang, K.J.; Zhang, X.Q. An experimental study on corona q–u curves under non–standard lightning impulses. *J. Electrost.* **2016**, *81*, 37–41. [CrossRef]
19. Ananea, Z.; Bayadia, A.; Haridb, N. A dynamic corona model for EMTP computation of multiple and non-standard impulses on transmission lines using a type-94 circuit component. *Electr. Power Syst. Res.* **2018**, *163*, 133–139. [CrossRef]
20. Mihailescu-Suliciu, M.; Suliciu, I. A rate type constitutive equation for the description of the corona effect. *IEEE Trans. Power App. Syst.* **1981**, *PAS-100*, 3681–3685. [CrossRef]
21. Zhang, X.Q. A realistic corona model for lightning transient studies. *Electr. Mach. Power Syst.* **1994**, *22*, 105–112.
22. Davis, R.; Cook, R.W.E.; Standring, W.G. The surge corona discharge. *Proc. IEE-Part C Monogr.* **1961**, *108*, 230–239. [CrossRef]
23. Maruvad, P.S.; Nguyen, D.H.; Hamadani-Zadeh, H. Studies on modeling corona attenuation of dynamic overvoltages. *IEEE Trans. Power Deliv.* **1989**, *4*, 1441–1449. [CrossRef]
24. Huang, W.G.; Semlyen, A. Computation of electromagnetic transients on three-phase transmission lines with corona and frequency dependent parameters. *IEEE Trans. Power Deliv.* **1987**, *2*, 887–898. [CrossRef]
25. Nawi, Z.M.; Ab Kadir, M.Z.A.; Azis, N.; Ahmad, W.F.; Rawi, I.N.; Ungku Amiruldm, A.U.; Nordin, F.H. Comparative analysis of ACSR and ACCC conductors on corona effect for lightning surge studies. In Proceedings of the 11th Asia-Pacific International Conference on Lightning, Hong Kong, China, 12–14 June 2019; pp. 1–5.
26. Dommel, H.W. *Electromagnetic Transients Program Theory Book*; BPA: Portland, OR, USA, 1995; pp. 3–19.
27. Sundaram, M.M.; Appadoo, D. Traditional salt-in-water electrolyte vs. water-in-salt electrolyte with binary metal oxide for symmetric supercapacitors: Capacitive vs. faradaic. *Dalton Trans.* **2020**, *49*, 11743–11755. [CrossRef]
28. Divakaran, A.M.; Hamilton, D.; Manjunatha, K.N.; Minakshi, M. Development and thermal analysis of reusable Li-ion battery module for future mobile and stationary applications. *Energies* **2020**, *13*, 1477. [CrossRef]
29. Razavi, S.E.; Arefi, A.; Ledwich, G.; Nourbakhsh, G.; Smith, D.B.; Minakshi, M. From load to net energy forecasting: Short-term residential forecasting for the blend of load and PV behind the meter. *IEEE Access* **2020**, *8*, 224343–224353. [CrossRef]
30. Beiza, J.; Hosseinian, S.H.; Vahidi, B. Multiphase transmission line modeling for voltage sag estimation. *Electr. Eng.* **2010**, *92*, 99–109. [CrossRef]
31. Hollman, J.A.; Marti, J.R. Step-by-step eigenvalue analysis with EMTP discrete-time solutions. *IEEE Trans. Power Syst.* **2010**, *25*, 1220–1231. [CrossRef]
32. Huang, K.J. Impulse Corona Characteristics and Their Influence on Transmission Line Transient Computation. Ph.D. Thesis, Beijing Jiaotong University, Beijing, China, April 2017.

Article

Physical Simulation of the Spectrum of Possible Electromagnetic Effects of Upward Streamer Discharges on Model Elements of Transmission Line Monitoring Systems Using Artificial Thunderstorm Cell

Nikolay Lysov *, Alexander Temnikov, Leonid Chernensky, Alexander Orlov, Olga Belova, Tatiana Kivshar, Dmitry Kovalev and Vadim Voevodin

Department of Electrophysics and High Voltage Technique, Moscow Power Engineering Institute, National Research University, 111250 Moscow, Russia; TemnikovAG@mpei.ru (A.T.); ChernenskyLL@mpei.ru (L.C.); orlovav@mpei.ru (A.O.); belovaos@mail.ru (O.B.); geratk@mail.ru (T.K.); kovalevdi@list.ru (D.K.); voevodinvv@mpei.ru (V.V.)
* Correspondence: streamer.corona@gmail.com; Tel.: +7-9168456253

Featured Application: The results of this work can be used to assess the reliability of electric power facilities' monitoring systems during thunderstorms. In addition to direct lightning strikes, the correct operation of monitoring systems can also be influenced by incomplete upward discharges, including those from nearby objects.

Abstract: The results of a physical simulation using negatively charged artificial thunderstorm cells to test the spectrum of possible electromagnetic effects of upward streamer discharges on the model elements of transmission line monitoring systems (sensor or antennas) are presented. Rod and elongated model elements with different electric field amplification coefficients are investigated. A generalization is made about the parameters of upward streamer current impulse and its electromagnetic effect on both kinds of model elements. A wavelet analysis of the upward streamer corona current impulse and of the signal simultaneously induced in the neighboring model element is conducted. A generalization of the spectral characteristics of the upward streamer current and of the signals induced by the electromagnetic radiation of the nearby impulse streamer corona on model elements is made. The reasons for super-high and ultra-high frequency ranges in the wavelet spectrum of the induced electromagnetic effect are discussed. The characteristic spectral ranges of the possible electromagnetic effect of upward streamer flash on the elements of transmission line monitoring systems are considered.

Keywords: artificial thunderstorm cell; lightning; upward streamer discharges; electromagnetic radiation spectrum; wavelet; transmission line monitoring system; model element; simulation

Citation: Lysov, N.; Temnikov, A.; Chernensky, L.; Orlov, A.; Belova, O.; Kivshar, T.; Kovalev, D.; Voevodin, V. Physical Simulation of the Spectrum of Possible Electromagnetic Effects of Upward Streamer Discharges on Model Elements of Transmission Line Monitoring Systems Using Artificial Thunderstorm Cell. *Appl. Sci.* **2021**, *11*, 8723. https://doi.org/10.3390/app11188723

Academic Editors: Massimo Brignone and Daniele Mestriner

Received: 9 August 2021
Accepted: 14 September 2021
Published: 18 September 2021

Publisher's Note: MDPI stays neutral with regard to jurisdictional claims in published maps and institutional affiliations.

1. Introduction

Software, computing complexes, and artificial intelligence algorithms are being increasingly introduced into power management systems, and include various remote monitoring systems for transmission lines. These systems use collected data to form control signals that make operational decisions [1–7]. At the same time, functional problems continue to appear during the use of digital technology and computing systems in the online monitoring of the air transmission lines (e.g., sensors of various kinds, analog–digital converters for the processing of recorded signals, and antenna and receiver-transmitting devices of different shapes and sizes [6–10]) under the influence of thunderclouds and lightning. Most often, these devices are rods, cylindrical, or flat. It is therefore necessary to ensure their electromagnetic compatibility [11–13]. Moreover, it is not entirely clear how these devices are affected by the different kinds of discharge phenomena that form on various design features of monitoring systems and/or the transmission lines while under the influence of thunderclouds and lightning (e.g., flashes of avalanche and streamer corona, ascending and downward leaders, the main discharge), or how the electromagnetic radiation they create will affect their functionality [12,14,15].

Those impacts are particularly dangerous, as they have frequencies close to the working frequencies of various elements and devices of the artificial intelligence system; this could be from hundreds of hertz to several gigahertz [6,8,12]. Sensors, receivers, communication systems, and in some cases, parts of the software-computing control complex of the transmission line monitoring system may be situated directly in the electric field area of thunderclouds and lightning discharge. In this case, exposure to electromagnetic radiation can occur in a wide frequency range, leading to interference, failures, distortions, false positives, and, accordingly, the disruption of normal functionality. Moreover, even the successful triggering of external lightning protection does not eliminate the possibility of an electrical objects' failure; this is a consequence of the impact of electromagnetic radiation of close lightning discharges in various frequency ranges at the different stages of its formation [14].

It is necessary to achieve a correct interpretation of the spectral characteristics of the electromagnetic radiation that affects the elements of the transmission line monitoring systems in the near field, and to determine their connection with the peculiarities of the formation of the lightning discharge between the thundercloud and the ground [16]. This requires investigation into the connection between the characteristic frequencies of the measured signal and the discharge processes taking place in the thundercloud; this must be conducted on objects on the surface of the earth and under a thundercloud, and between the thundercloud and the ground during the formation of lightning discharge [16–20].

The use of artificial thunderstorm cells of negative polarity makes it possible to physically simulate and investigate the characteristics of various types of electrical discharges and the electromagnetic radiation they create. This can be formed on the model elements of the monitoring systems of transmission lines, or be imposed on them during close lightning strikes. The purpose of this work is to physically model (using artificial thunderstorm cells) the spectrum of the possible direct and induced electromagnetic impact of upward streamer discharges in order to determine their influence on the functionality of the intelligent systems that monitor the air transmission lines. These discharges may form on the transmission line intelligent monitoring systems (receiving-transmission devices of different kinds), as well as on the adjacent, grounded structures of the transmission line.

2. Experimental Schemes

This research was performed using equipment from the core shared research facilities, namely the high-voltage research complex of the National Research University, the Moscow Power Engineering Institute, which allows the creation of artificial storm cells of negative polarity with a potential of up to 1.5 MV [21]. As a result, a strong electric field appears in the gap between the artificial thunderstorm cell of negative polarity and the grounded plane, which develops all the forms of spark discharge that are characteristic of a thunderstorm, including the leader and main stage.

The measuring units of monitoring systems are usually located on transmission towers and phase wires. The measuring unit includes sensors for measuring main parameters, a processor module, and a data transmission system. Depending on their functional purpose, monitoring systems can use various types of sensors and transmitters (transceivers) of a rod or elongated type with sizes ranging from several centimeters to tens of centimeters. Sensors can have an almost spherical or cylindrical shape, or have a complex shape with protruding rod elements [5,6,8]. Since the structural elements of the transmission tower, as well as the phase and grounded wires, are also close in shape to a rod or cylindrical form, the following two experimental schemes with rod or cylindrical electrodes were chosen for the physical modeling used during the research.

Two experimental schemes were used for a physical simulation of the spectrum. The first scheme simulated upward streamer discharges formed from the rod sensors (or antenna devices) or the effects of the electromagnetic radiation of the upward streamer discharges, formed on the rod elements of the transmission line design on the nearby model sensors (Figure 1). The second scheme simulated the analogous situation, but for the case of sensors (or antenna devices) of the cylinder type and of limited length, and the phase and ground wires (Figure 2). The distance between two grounded electrodes (one simulating the element from which an ascending streamer discharge is formed, and the other simulating the element on which the signal is induced by this discharge) was in the range of 15 to 30 cm. In the simulation, a monitoring system element and the places of formation of ascending streamer discharges on an overhead power transmission line were situated relatively close together.

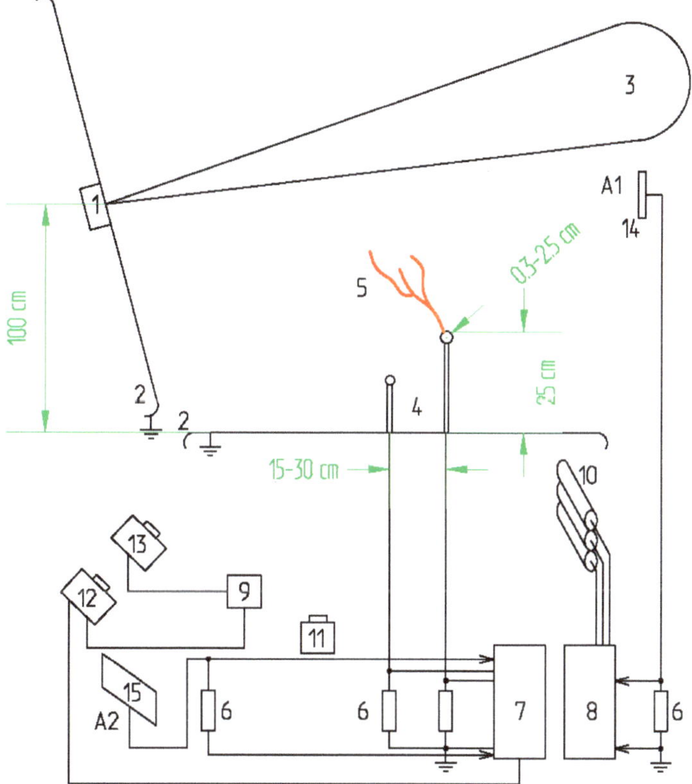

Figure 1. First scheme of experimental and measurement setup: 1—charged aerosol generator; 2—grounded electrostatic screens; 3—artificial thunderstorm cell; 4—rod electrodes; 5—upward streamer discharge; 6—shunts; 7,8—digital oscilloscope; 9—trigger generator; 10—system of photomultipliers; 11—digital photo camera; 12—photomultiplier; 13—CCD-camera; 14, 15—flat antennas.

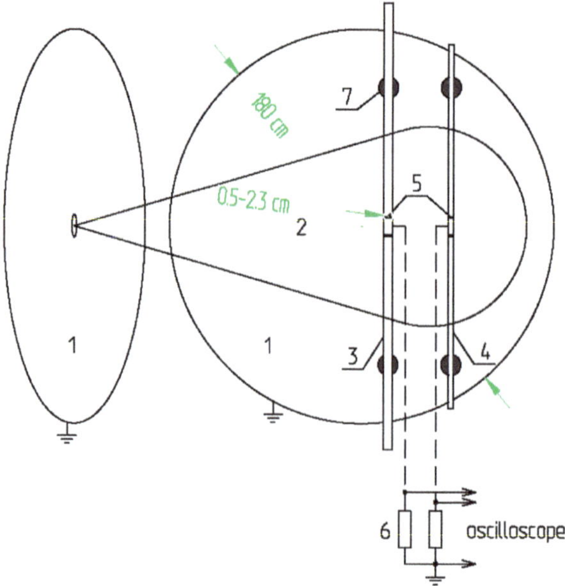

Figure 2. Second scheme of experimental and measurement setup: 1—grounded electrostatic screens; 2—artificial thunderstorm cell; 3,4—cylinder electrodes; 5—isolated elongated elements; 6—shunts; 7—insulators.

The conditions for the formation (the occurrence and subsequent development) of ascending streamer discharges originating from the grounded elements of monitoring systems in an external electric field created by a thundercloud and/or a descending lightning leader significantly depend on the nature of the distribution of the local electric field in the area near them [22]. Therefore, when conducting experimental studies, the radii of the vertices of rod model objects and the radii of model cylindrical objects varied in range from 0.3 to 2.5 cm for rod elements, and from 0.5 to 2.3 cm for extended elements. The height of the electrodes varied from 15 to 37 cm. Model rod electrodes were made of brass or aluminum; model cylindrical electrodes (tubes) were made of brass, aluminum, or steel. Elements of sensors, transmission towers, and phase and grounded wires were also made of these materials. As a result, the experiment simulated the formation of ascending discharge phenomena from elements of cyber-physical objects and systems with significantly different electric field amplification coefficients under the influence of atmospheric electricity and lightning. To analyze the influence this factor had on the experimental results, all model elements were divided into three groups according to the electric field amplification coefficient: group I had an amplification coefficient < 10; group II had an amplification coefficient < 25; group III had an amplification coefficient > 25.

Examples of the formation of upward streamer discharges from the grounded rods and elongated model elements under the negative polarity artificial thunderstorm cell are shown in Figures 3 and 4, respectively.

Figure 3. Upward streamer corona flash from the grounded rod model elements.

Figure 4. Upward streamer discharges from the grounded cylinder model element.

The characteristic oscillograms of the current impulse of powerful streamer corona flash and the corresponding induced electromagnetic effects (induced current) on the close rod or elongated model element are shown in Figure 5. The discharge current was registered using low-inductance shunts (6, Figures 1 and 2). The current induced by the discharge was registered using flat antennas (A1 and A2, Figure 1). Both signals were recorded by digital oscilloscopes Tektronix DPO7254 and Tektronix TDS3054C (Tektronix, Inc., Beaverton, OR, USA).

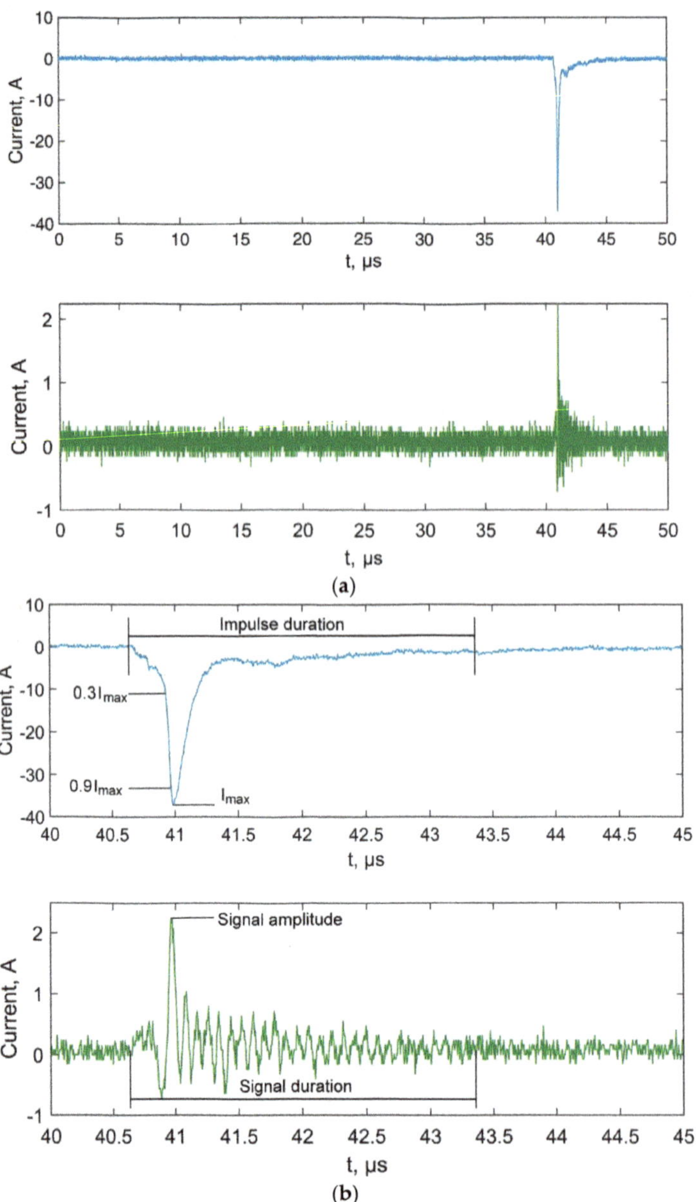

Figure 5. (**a**) Oscillograms of the upward streamer discharge current (upper) and induced electromagnetic effects (bottom). (**b**) Focused oscillograms of the formation of the upward streamer discharge current (upper) and induced electromagnetic effects (bottom).

For the impulse streamer corona flash of the upward discharge current amplitude, maximal current rise velocity, flowing charge, and impulse duration were determined. For induced electromagnetic effects on the signal amplitude of the nearby model elements, the duration of the induced signal was determined. The maximum value of the current pulse was described as the current amplitude (Imax). The time between the start and end of the impulse was described as the impulse duration. The start time of the pulse was decided based on the first time it crossed the zero value; similarly, the end time was decided based on the time of the first signal zero value following Imax. The maximal current rise velocity was specified as the ratio of the difference between 0.9 and 0.3 Imax of the signal duration between the points corresponding to 0.9 and 0.3 Imax (Figure 5b). The flowing charge was estimated by the integration of the current pulse from the pulse start time to the end time.

The spectral characteristics of the discharge current and the induced signals were determined based on a wavelet analysis, using the specially created program and the "Mexican Hat" wavelet [23,24]. In mathematics, a wavelet series is a representation of a square-integrable (real- or complex-valued) function by a certain orthonormal series generated by a wavelet. The fundamental idea behind wavelet transforms is that the transformation should only allow changes in the time extension, but not the shape. While the Fourier transform creates a representation of the signal in the frequency domain, the wavelet transform creates a representation of the signal in both the time and the frequency domain, thereby allowing efficient access to localized information about the signal. The upper level of the characteristic frequency, maximal intensity, and frequency of the maximal intensity in the wavelet spectrum was found. The characteristic wavelet spectrum for the currents and induced signal presented in Figure 5 are shown in Figures 6 and 7, respectively.

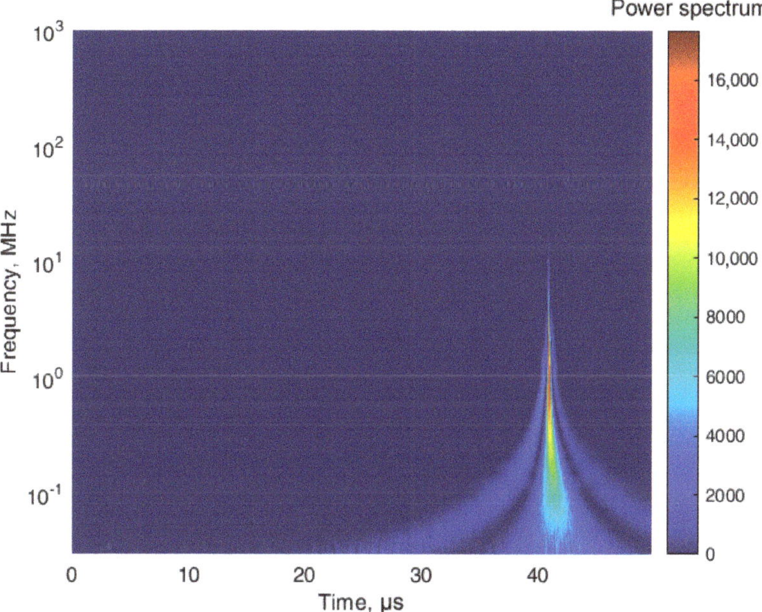

Figure 6. Wavelet spectrum of upward streamer corona current impulse.

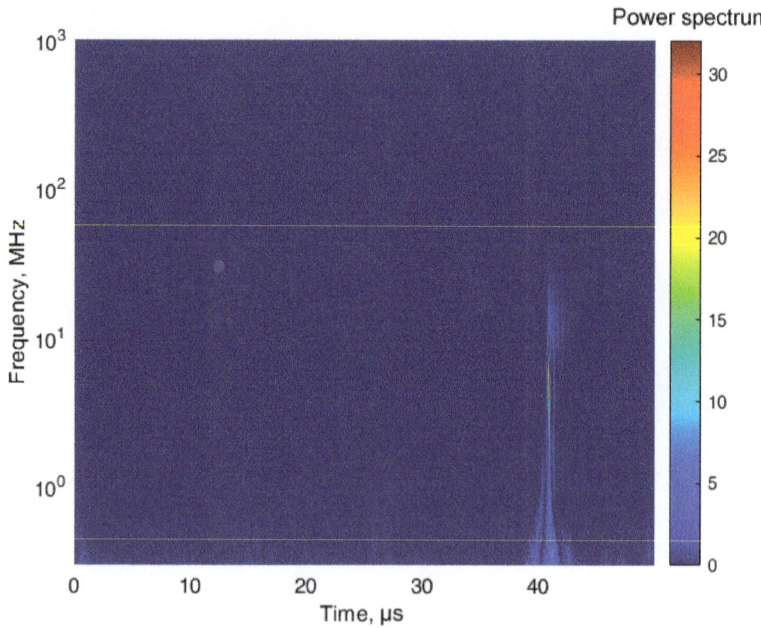

Figure 7. Wavelet spectrum of the signal induced in the nearby model element by the upward streamer corona flash.

3. Analysis of Results and Discussion

During the study, 530 experimental attempts were performed and processed, 222 of which were performed using the first experimental scheme, and 308 using the second.

The processed experimental results showed that the characteristics of the current pulse of the upward streamer corona with model rods and elongated elements depended on the characteristics of the electric field (group of amplification coefficient) near such objects (Tables 1 and 2). The impulse current of the streamer flash from the model elements with relatively low amplification coefficients (group I) showed average higher current amplitudes and charges, and less impulse duration. However, higher values of the maximal current rise velocity were observed for the model elements with amplification coefficients ranging from 10 to 25 (group II). For all groups of model amplification coefficients, the duration of the current impulse and flowing charge was on average higher for elongated model elements than for rods.

Table 1. Characteristics of the current impulse of streamer corona on the rod model elements (average values).

Amplification Coefficient	Group I	Group II	Group III
Current amplitude, A	8.45	5.11	3.26
Maximal current rise velocity, A/ns	0.13	0.17	0.14
Impulse duration, µs	2.65	5.24	2.98
Flowing charge, µC	4.38	4.17	1.96

It should be noted that for all groups of model element amplification coefficients, the characteristics of the signals induced in the elongated model elements by the neighboring upward streamer corona flash showed higher values than those of the rod model elements (Table 3).

Table 2. Characteristics of the current impulse of streamer corona on the cylinder model elements (average values).

Amplification Coefficient	Group I	Group II	Group III
Current amplitude, A	8.89	3.04	2.19
Maximal current rise velocity, A/ns	0.10	0.13	0.12
Impulse duration, μs	4.12	7.08	9.23
Flowing charge, μC	6.54	5.09	4.12

Table 3. Characteristics of the signals induced with nearby impulse streamer corona on the model elements (average values).

Source of Effect	Streamer Corona (Rod Model Element)		Streamer Corona (Elongated Model Element)	
Amplification Coefficient	Signal Amplitude, A	Signal Duration, μs	Signal Amplitude, A	Signal Duration, μs
Group I	0.94	1.06	1.18	1.12
Group II	1.36	0.31	1.41	0.52
Group III	1.02	0.69	1.22	0.97

The wavelet analysis showed that the spectral characteristics (upper frequency and frequency corresponding to maximal intensity) of a current impulse of an upward streamer flash had average values 1.5–9.0 times higher for the case of the rod model elements than for elongated elements for all the groups of model amplification coefficients (Tables 4 and 5). In some cases, an upper frequency ranging from 500 MHz to 1 GHz appeared in the wavelet spectrum of the current impulse of the streamer corona from the rod model elements related to group III (a higher amplification coefficient).

Table 4. Spectral characteristics of the current impulse of streamer corona on the rod model elements (average values).

Amplification Coefficient	Group I	Group II	Group III
Upper level of the characteristic frequency, MHz	9.9	13.2	83.5
Frequency of the maximal intensity, MHz	0.9	1.1	5.8
Maximal intensity, A2	2300	900	450

Table 5. Spectral characteristics of the current impulse of streamer corona on the elongated model elements (average values).

Amplification Coefficient	Group I	Group II	Group III
Upper level of the characteristic frequency, MHz	6.8	5.2	9.5
Frequency of the maximal intensity, MHz	0.5	0.4	1.8
Maximal intensity, A2	3600	800	200

A generalization of the spectral characteristics of the signals induced by the electromagnetic radiation of the nearby impulse streamer corona flash on model elements of different types showed the close values of all frequency parameters inside every group of the electric amplification coefficient for rod and elongated model elements (Tables 6 and 7). The maximal values of the upper level of the characteristic frequency in the wavelet spectrum of the induced signal (the electromagnetic radiation of upward streamers was sometimes in the range of 1.0–1.5 GHz) were discovered in group II (an amplification coefficient of 10–25). Group III (amplification coefficient > 25) showed similar, but slightly lower, values.

Table 6. Spectral characteristics of the signals induced with nearby impulse streamer corona on the rod model elements (average values).

Amplification Coefficient	Group I	Group II	Group III
Upper level of the characteristic frequency, MHz	111	795	397
Frequency of the maximal intensity, MHz	10	76	27
Maximal intensity, A2	10	14	9

Table 7. Spectral characteristics of the signals induced with nearby impulse streamer corona on the elongated model elements (average values).

Amplification Coefficient	Group I	Group II	Group III
Upper level of the characteristic frequency, MHz	123	809	528
Frequency of the maximal intensity, MHz	8	68	18
Maximal intensity, A2	12	10	9

Presumably, such high frequencies appear in the spectrum of the electromagnetic radiation of the upward streamer discharges due to their development in a highly divergent electric field. This leads to the rapid rise in the discharge current. As seen in Tables 1 and 2, the maximal current rise velocity was observed on model elements from group II.

Thus, it is suggested that the formation of electromagnetic radiation of the corona discharge in an avalanche and streamer form is presumably one of the key mechanisms of the electromagnetic influence on the discharge phenomena in artificial and natural storm cells and clouds that are located in the vicinity of model elements of cyber-physical objects, and systems in the super-high and ultra-high frequency range [15,25–27]. According to [26], the electromagnetic radiation of avalanches formed from rods and long (cylindrical) electrodes and in the head of streamers in various large electric fields have an electromagnetic radiation that can be clearly expressed in the range of 4.0–5.0 MHz to 0.9–1.0 GHz, from 1.0–2.0 MHz to 0.5–0.6 GHz, and from 0.5–1.0 MHz to 8–10 GHz. Another source of such ultra-high electromagnetic radiation of the streamer flash, as proposed in [27], could be the result of the streamers colliding inside a streamer zone. Both of these electromagnetic radiation formation mechanisms correlate to the characteristic frequency ranges of the wavelet spectrums of signals, and are guided on model rods and long elements of the neighboring flash of the streamer corona; these develop from another model element belonging to group II and group III, containing large force-strength factors of the electric field. Electromagnetic radiation with frequencies exceeding 1 GHz during streamer flashes in the electric field of a cloud charged with water drops was registered in [28] as well.

The possible electromagnetic effects of upward streamer flash on elements of transmission line monitoring systems in the determined characteristic spectral ranges are speculated. If upward streamer discharges form on the sensors (especially the rod kind, and on those included in group III) of the monitoring system, frequencies that are in the spectrum of a current impulse (up to dozens or hundreds of megahertz) will be in the working frequency range of the sensor. This could lead to sensor failure, including errors in the digital processing of data by the analog–digital converters [11–13,29].

Similar failures may also be encountered when the electromagnetic radiation of streamer corona flash originates near the monitoring, diagnostic, and control systems that affect its elements (e.g., sensors). Failures may also occur if analog–digital converters with working frequencies from several hundred kilohertz to several gigahertz are used for subsequent digital processing of the measuring information [5,6,8,11]. The electromagnetic radiation of the streamer corona flash ranging from hundreds of megahertz to gigahertz could impact nearby receiving/transmitting devices, and result in disruption, distortion, or a loss in informational transmission [8,11,12].

4. Conclusions

Physical modeling (using an artificial thunderstorm cell) of the possible influence of upward streamer discharges from the rod or elongated model elements of a power transmission line monitoring systems on their operation in the electric field of a thunderstorm cloud and/or lightning produced many results. Firstly, upward streamer discharges that formed on the monitoring systems model elements (i.e., sensors, and receiving and transmitting devices) can affect the functioning of these systems due to the fact that in the spectrum of their currents, there are frequencies of up to tens of megahertz, which is close to the operating frequencies of power transmission line monitoring systems. Secondly, upward streamer discharges on the model elements of power transmission lines can also induce signals in the neighboring rod or elongated elements of the monitoring systems. These signals are dangerous for the monitoring systems' operation, since the discharge spectrum contains frequencies of tens to hundreds of megahertz (up to units of gigahertz).

It was found that the parameters of the current pulse of upward streamer discharges, the signals that they induce in the neighboring elements, and their spectral characteristics depend on the electric field distribution near the model rod or elongated element (electric field amplification factor). The highest frequencies in the spectrum of the current pulse of the streamer corona flash are typical for the model elements with a high electric field amplification coefficient of >25. The highest frequencies in the spectrum of the signal induced in the neighboring element by the upward streamer discharges are typical for model elements with an amplification coefficient of 10–25. Thus, both the frequency ranges in the spectrum of the current pulse of upward streamer discharges formed from the model elements of sensors and receiving-transmitting devices of the power transmission line monitoring system and the frequency ranges in the spectrum of signals induced by close upward streamer discharges on these elements may appear close to the operating frequency ranges of analog-to-digital converters of sensors and/or devices for transmitting data in monitoring systems. This can lead to failures in monitoring systems operation, false alarms, incorrect transmission of information, and, as a result, create significant risks for the functionality of these systems during thunderstorms.

Author Contributions: Project administration and methodology, A.T.; software, L.C. and O.B.; visualization, T.K.; investigation, A.O. and N.L.; writing—original draft preparation, V.V.; funding acquisition, D.K. All authors have read and agreed to the published version of the manuscript.

Funding: This study conducted by Moscow Power Engineering Institute was financially supported by the Ministry of Science and Higher Education of the Russian Federation (project No. FSWF-2020-0019).

Institutional Review Board Statement: Not applicable.

Informed Consent Statement: Not applicable.

Data Availability Statement: Not applicable.

Conflicts of Interest: The authors declare no conflict of interest.

References

1. Hu, Y.; Liu, K. *Inspection and Monitoring Technologies of Transmission Lines with Remote Sensing*; Academic Press: Cambridge, MA, USA, 2017.
2. Li, S.; Li, J. Condition monitoring and diagnosis of power equipment: Review and prospective. *High Volt.* **2017**, *2*, 82–91. [CrossRef]
3. Deng, C.-J. Challenges and Prospects of Power Transmission Line Intelligent Monitoring Technology. *Am. Res. J. Comput. Sci. Inf. Technol.* **2019**, *4*, 1–11.
4. Zhirui, L.; Shengsuo, N.; Nan, J. Current Status and Development Trend of AC Transmission Line Parameter Measurement. *Autom. Electr. Power Syst.* **2017**, *41*, 181–191.
5. Liu, Y.; Yin, H.; Wu, T. Transmission Line on-line Monitoring System Based on Ethernet and McWiLL. In Proceedings of the International Conference on Logistics Engineering, Management and Computer Science (LEMCS 2015), Shenyang, China, 26–28 June 2015; pp. 680–683.
6. Working Group on Monitoring & Rating of Subcommittee 15.11 on Overhead Lines. Real-Time Overhead Transmission-Line Monitoring for Dynamic Rating. *IEEE Trans. Power Deliv.* **2016**, *31*, 921–927. [CrossRef]

7.   Xing, Z.; Cui, W.C.; Liu, R.; Zheng, Z. Design and Application of Transmission Line Intelligent Monitoring System. In Proceedings of the 2020 International Conference on Energy, Environment and Bioengineering (ICEEB 2020), Xi'an, China, 7–9 August 2020; Volume 185. [CrossRef]
8.   McCall, J.C.; Spillane, P.; Lindsey, K. Determining Crossing Conductor Clearance Using Line-Mounted LiDAR. In Proceedings of the CIGRE US National Committee 2015 Grid of the Future Symposium, Paris, France, 12 October 2015.
9.   Wydra, M.; Kubaczynski, P.; Mazur, K.; Ksiezopolski, B. Time-Aware Monitoring of Overhead Transmission Line Sag and Temperature with LoRa Communication. *Energies* **2019**, *12*, 505. [CrossRef]
10.   Chen, H.; Qian, Z.; Liu, C.; Wu, J.; Li, W.; He, X. Time-Multiplexed Self-Powered Wireless Current Sensor for Power Transmission Lines. *Energies* **2021**, *14*, 1561. [CrossRef]
11.   Hoole, P.R.P.; Sharip, M.R.M.; Fisher, J.; Pirapaharan, K.; Othman, A.K.H.; Julai, N.; Rufus, S.A.; Sahrani, S.; Hoole, S.R.H. Lightning Protection of Aircraft, Power Systems and Houses Containing IT Network Electronics. *J. Telecommun. Electron. Comput. Eng.* **2017**, *9*, 3–10.
12.   Ahmad, M.R.; Esa, M.R.M.; Cooray, V.; Dutkiewicz, E. Interference from cloud-to-ground and cloud flashes in wireless communication system. *Electr. Power Syst. Res.* **2014**, *113*, 237–246. [CrossRef]
13.   Borisov, R.K.; Zhulikov, S.S.; Koshelev, M.A.; Maksimov, B.K.; Mirzabekyan, G.Z.; Turchaninova, Y.S.; Khrenov, S.I. A Computer-aided design system for protecting substations and overhead power lines from lightning. *Russ. Electr. Eng.* **2019**, *90*, 86–91. [CrossRef]
14.   Cooray, V. *Lightning Electromagnetics*; IET Publishing: London, UK, 2012.
15.   Cooray, V.; Cooray, G. Electromagnetic fields of accelerating charges: Applications in lightning protection. *Electr. Power Syst. Res.* **2017**, *145*, 234–247. [CrossRef]
16.   Nag, A.; Murphy, M.J.; Schulz, W.; Cummins, K.L. Lightning location systems: Insights on characteristics and validation technique. *Earth Space Sci.* **2015**, *2*, 65–93. [CrossRef]
17.   Chen, M.; Du, Y.; Burnett, J.; Dong, W. The electromagnetic radiation from lightning in the interval of 10 kHz to 100 MHz. In Proceedings of the 12th International Conference on Atmospheric Electricity, Versailles, France, 9–14 June 2003.
18.   Makela, J. *Electromagnetic Signatures of Lightning near the HF Frequency Band*; Finnish Meteorological Institute: Helsinki, Finland, 2009.
19.   Dong, W.; Liu, H. Observation of compact intracloud discharges using VHF broadband interferometer. In Proceedings of the 2012 International Conference on Lightning Protection (ICLP), Vienna, Austria, 2–7 September 2012.
20.   Rakov, V.A. Electromagnetic Methods of Lightning Detection. *Surv. Geophys.* **2013**, *34*, 731–753. [CrossRef]
21.   Temnikov, A.G. Using of artificial clouds of charged water aerosol for investigations of physics of lightning and lightning protection. In Proceedings of the 2012 International Conference on Lightning Protection (ICLP), Vienna, Austria, 2–7 September 2012. [CrossRef]
22.   Bazelyan, E.M.; Raizer, Y.P. *Lightning Physics and Lightning Protection*; IoP Publishing: Bristol, UK; New York, NY, USA, 2000.
23.   Esa, M.R.M.; Ahmad, M.R.; Cooray, V. Wavelet analysis of the first electric field pulse of lightning flashes in Sweden. *J. Atmos. Res.* **2014**, *138*, 253–267. [CrossRef]
24.   Temnikov, A.G.; Chernensky, L.L.; Belova, O.S.; Orlov, A.V.; Zimin, A.S. Spectral characteristics of discharges from artificial charged aerosol cloud. In Proceedings of the 2014 International Conference on Lightning Protection (ICLP), Shanghai, China, 11–18 October 2014.
25.   Cooray, V.; Cooray, G. The Electromagnetic Fields of an Accelerating Charge: Applications in Lightning Return-Stroke Models. *IEEE Trans. Electromagn. Compat.* **2010**, *52*, 944–955. [CrossRef]
26.   Cooray, V.V.; Cooray, G. Electromagnetic radiation field of an electron avalanche. *Atmos. Res.* **2012**, *117*, 18–27. [CrossRef]
27.   Shi, F.; Liu, N.; Dwyer, J.R.; Ihaddadene, K.M.A. VHF and UHF electromagnetic radiation produced by streamers in lightning. *Geophys. Res. Lett.* **2019**, *46*, 443–451. [CrossRef]
28.   Gushchin, M.E.; Korobkov, S.V.; Zudin, I.Y.; Nikolenko, A.S.; Mikryukov, P.A.; Syssoev, V.S.; Sukharevsky, D.I.; Orlov, A.I.; Naumova, M.Y.; Kuznetsov, Y.A.; et al. Nanosecond electromagnetic pulses generated by electric discharges: Observation with clouds of charged water droplets and implications for lightning. *Geophys. Res. Lett.* **2021**, *48*, e2020GL092108. [CrossRef]
29.   Judge, M.A.; Manzoor, A.; Ahmed, F.; Kazmi, S.; Khan, Z.A.; Qasim, U.; Javaid, N. Monitoring of Power Transmission Lines Through Wireless Sensor Networks in Smart Grid. In Proceedings of the 11th International Conference on Innovative Mobile and Internet Services in Ubiquitous Computing, Torino, Italy, 28–30 June 2017.

Article

Impact of Grounding Modeling on Lightning-Induced Voltages Evaluation in Distribution Lines

Daniele Mestriner [1,†], Rodolfo Antônio Ribeiro de Moura [2,*,†], Renato Procopio [1,†] and Marco Aurélio de Oliveira Schroeder [2,†]

[1] Naval, ICT and Electrical Engineering Department (DITEN), University of Genoa, Via Opera Pia 11a, 16145 Genoa, Italy; daniele.mestriner@edu.unige.it (D.M.); renato.procopio@unige.it (R.P.)

[2] Electrical Engineering Department, Federal University of São João del-Rei, Praça Frei Orlando, 170-Centro, 36307-352 São João del-Rei, MG, Brazil; schroeder@ufsj.edu.br

[*] Correspondence: moura@ufsj.edu.br; Tel.: +55-32-99102-0091

[†] These authors contributed equally to this work.

Abstract: Lightning-induced voltages are one of the main causes of shutdown in distribution lines. In this work, attention is focused on the effects of wideband modeling of electric grounding in the overvoltage calculation along insulator strings due to indirect lightning strikes. This study is done directly in the time-domain with the grounding being represented with an equivalent circuit accounting for its dynamics. Results show that the adoption of commonly adopted simplified grounding models, such as low-frequency resistance, may lead to an underestimation of the overvoltage. According to the results, differences in the order of 25% can be found in some studied cases.

Keywords: distribution lines; lightning-induced overvoltages; grounding modeling; soil resistivity

 check for **updates**

Citation: Mestriner, D.; Ribeiro de Moura, R.A.; Procopio, R.; de Oliveira Schroeder, M.A. Evaluation of the Impact of Grounding Modeling on Lightning-Induced in Distribution Lines. *Appl. Sci.* **2021**, *11*, 2931. https://doi.org/10.3390/app11072931

Academic Editor: Matti Lehtonen

Received: 26 January 2021
Accepted: 22 March 2021
Published: 25 March 2021

1. Introduction

Transmission and Distribution Systems are highly affected and damaged by direct and indirect lightning events. Direct events occur when lightning directly strikes the line; such events are hazardous but rare and are typically studied and analyzed in Transmission System (TS). On the other hand, indirect events occur when lightning strikes the ground in the proximity of a power system; these events are much more frequent with respect to direct ones, but the overall voltage induced in the power system is usually much lower. For this reason, indirect events are not of interest for TS since the induced voltages are generally lower than the line Critical FlashOver voltage (CFO), but they are vital when dealing with Distribution Systems (DS), which are characterized by a low CFO.

Most works address lightning-induced voltages in DS model electric grounding as a constant value resistance R_{LF} [1–16]. This parameter is associated with a low-frequency behavior, i.e., disregarding its electromagnetic dynamic. Therefore, this low-frequency grounding resistance cannot reproduce the reactive (inductive and capacitive) and electromagnetic wave propagation effects (attenuation and distortion), prominent in the high-frequency range related to the voltage and current wavefronts. Additionally, the determination of overvoltage on TS, due to direct lightning, is highly sensible on the electromagnetic modeling of the electrical grounding [17].

Given the above, this work presents an evaluation of the impact of grounding modeling on lightning-induced voltage. Thus, the main original contribution of this paper is to include, in the time domain type simulations, an equivalent electric circuit that reproduces the complete frequency response of grounding, with full inclusion of the aforementioned effects. The Hybrid Electromagnetic Model (HEM) is used to determine the wideband grounding frequency response $Z(\omega)$ [18,19]. To implement the $Z(\omega)$ in silico, the Vector Fitting (VF) technique is applied to generate an equivalent electric circuit that is easily inserted in EMT-type software [20,21]. In the following, the grounding circuit will be implemented in the software developed in [22]. In this paper, as commonly proposed in the IEEE Standard [5], the coupling between the tower and the lightning channel and the coupling between the lightning channel and the grounding electrodes are neglected.

The results illustrate that the induced voltages, considering the grounding modeled via R_{LF} are quite different from those results using $Z(\omega)$, with perceptual differences reaching values of around 25%. It is noticeable that the differences increase with the soil resistivity and with the point of occurrence of the lightning (lightning striking closer to the DS increase the perceptual differences) for both first and subsequent return strokes. The paper is organized as follows: Sections 2–4 show the lightning field-to-line coupling problem equations, the tower and the grounding modeling, respectively; while Sections 5 and 6 present the test cases and the results. Section 7 is dedicated to the conclusions.

2. Induced-Lightning Modeling

The lightning-induced voltages occurring in a DS are here evaluated, recalling the procedure presented in [22,23]. This procedure is usually divided into two steps: (i) the ElectroMagnetic (EM) fields computation and (ii) the field-to-line coupling.

2.1. EM Fields Computation

The EM fields are computed analytically considering the approach proposed in [24] and validated in [25]. The method requires as input the knowledge of the channel-base current, the return stroke height and the return stroke velocity. It can be applied both to perfect electric conductor ground and soil characterized by a finite conductivity. The only assumption required is the Transmission Line model for the attenuation of the current along the channel. The main advantage of this approach consists of the possibility of dealing with analytical formulas, which guarantee a fast solution and a low computational effort.

2.2. Field-to-Line Coupling

The field-to-line coupling computation is obtained considering the well-known Agrawal model [26], which is here presented in its extended version taking into account the presence of a finite-conducting ground and a multi-conductor line.

$$\begin{cases} \frac{\partial V_i^s}{\partial x}(x,t) + \sum_{j=1}^{M} L_{ij}\frac{\partial I_j}{\partial t}(x,t) + V_i^g(x,t) = E_{inc,x,i}(x,t) \\ \frac{\partial I_i}{\partial x}(x,t) + \sum_{j=1}^{M} C_{ij}\frac{\partial V_j^s}{\partial t}(x,t) = 0 \end{cases} \tag{1}$$

with

$$V_i^g(x,t) = \int_0^t \xi_g^i(t-s)\frac{\partial I_i}{\partial s}(x,s)ds \tag{2}$$

where $V_i^s(x,t)$, $I_i(x,t)$ and $E_{inc,x,i}(x,t)$ are the scattered voltage, the current and the tangential component of the exciting electric field (computed in the previous subsection) on the ith conductor at distance x from the beginning of the line. As expressed in Equation (1), the knowledge of the inductance and capacitance matrices (L and C) is required. Please note that ξ_i^g is the time-domain expression for the ground impedance [27].

The total voltage occurring on the i-th conductor at the point x can be then expressed as the sum of the scattered voltage and the incident voltage, whose value depends on the vertical electric field (computed in the previous subsection).

The proposed methodology is adapted to an EMT-type software (in this framework Simulink-Simscape is used), through the finite-difference time-domain (FDTD) technique. In this case, a second-order scheme is adopted with $dt = 10$ ns and $dx = 9$ m, which satisfies the well-known Courant stability condition. Further details can be found in [22].

3. Tower Modeling

The modeling of the tower is usually neglected in lightning-induced voltages studies. However, in this framework, the tower is included in the model according to [28,29] and is modeled as a lossless transmission line, whose characteristic impedance is:

$$Z_c = 60 \left[\ln \left(\sqrt{2} \frac{2h}{r} \right) - 1 + \frac{r}{4h} + \left(\frac{r}{4h} \right)^2 \right] \tag{3}$$

h being the tower height and r the tower radius.

4. Electrical Grounding Modeling

In this paper, the grounding transient behavior is modeled by HEM [18,19]. This model is an electromagnetic computational method developed for the numerical solution of lightning problems and, according to CIGRÉ [30], it is classified as a hybrid electromagnetic-circuit approach. The main motivations for using HEM are as follows: (i) it is accurate and flexible, i.e., it can be used in different types of grounding configuration; (ii) its results have been extensively validated experimentally, such as measurements in TS [18,19], horizontal electrodes [18,31], vertical rods [31,32], and typical substation grounding grids [33] and (iii) it is faster than traditional full-wave methods (without losing accuracy). It is worth mentioning that the usage of HEM has increased significantly recently [30,34,35].

Basically, HEM consists of subdividing the actual system (in this case, electrical grounding) into N small conductive cylindrical segments and, for each segment, the electromagnetic theory is applied. After that, by using the circuit theory, it is possible to obtain a matrix system that computes the wideband response of the electrical grounding. For the sake of clarity, we present a brief overview of HEM below. More details about HEM are described in [18,19].

It is worth commenting on the fact that HEM corresponds to an electromagnetic model developed in the frequency domain. Thus, it is necessary first to determine the frequency spectrum (depending on the phenomenon of interest). After that, the electrical grounding is divided into N segments, where the length of each segment is equal to 10 times its radius (thin wire approximation). A discussion about segmentation length is presented in [36].

Each segment is considered a source of two currents, one longitudinal that flows along the electrode (I_L) and another transversal that flows from the electrode to the surrounding soil (I_T). It is worth noting that I_L generates a non-conservative electric field and I_T a conservative one. With the aid of the magnetic vector and electric scalar potentials, both voltage drops (ΔV) and electric potentials (V) in each pair of segments (transmitter and receiver) are determined. Additionally, double integral equations are established for ΔV and V. These integrals depend on the frequency, geometry, soil parameters and I_T and I_L distributions. However, the distributions of I_T and I_L are not known and are integrands of the integrals. From this point on, the Method of Moments (MoM) is applied to solve these integral equations [37]. The effect of the air-soil interface is included using the method of images, similar to [38,39].

The I_T and I_L distributions considered in this paper are of the piecewise-constant function type [18,19]. MoM makes it possible to transform integral equations into algebraic ones, the solution of which allows determining all the quantities of interest (in the frequency domain), such as I_T, I_L, ΔV and V distributions; transverse (capacitive and conductive couplings) and longitudinal (resistive and inductive couplings) impedances (self and mutual); electromagnetic field; harmonic grounding impedance ($Z(\omega)$); low-frequency grounding resistance (R_{LF}), etc.

Also, it has been documented in the literature, over almost one hundred years [40], that the soil is a dispersive medium, i.e., the response is not instantaneous. Several researchers have presented a numerical solution to consider this dispersivity in the frequency domain, such as [31,40–49]. According to [50], for the values of conductivity considered in this paper, the frequency-dependent soil electric parameters can be neglected in the computation of the EM fields that illuminate the line. On the other hand, according to [17], the impact of considering the frequency-dependence of the soil in the grounding modeling generates sensible differences. Thus, in this paper, the frequency-dependence of the soil parameters is considered in the grounding modeling. Use is made by the formulations proposed in [31], since a CIGRE Brochure has suggested them [51]. Equations (4) and (5) illustrate the formulation.

$$\sigma(f) = \sigma_0 + \sigma_0 h(\sigma_0)\left(f \times 10^{-6}\right)^{\gamma} \tag{4}$$

$$\epsilon_r(f) = \epsilon_{r\infty} + \frac{\tan(\pi\gamma/2) \times 10^{-3}}{2\pi\epsilon_0(10^6)^{\gamma}}\sigma_0 h(\sigma_0) f^{(\gamma-1)} \tag{5}$$

where $\sigma(f)$ is the frequency-dependent soil conductivity (in mS/m), σ_0 is the low-frequency soil conductivity (in mS/m), $\epsilon_r(f)$ is the frequency-dependent soil permittivity and ϵ_0 is the vacuum permittivity. To obtain mean results (more details about it in [31]), one should use $h(\sigma_0) = 1.26\sigma_0^{-0.73}$, $\gamma = 0.54$ and $\epsilon_{r\infty} = 12$.

As specified before, the calculation of induced voltages is performed directly in the time domain; however, $Z(\omega)$ is a frequency domain quantity. Thus, the well-known Vector Fitting (VF) approach is used for fitting the calculated frequency domain grounding response with rational function approximations [20]. The passivity is enforced by perturbation [21].

Finally, based on the obtained rational function, it is possible to synthesize an electric network that can be promptly included in the time-domain simulation. It is important to note that this electric circuit generates the same frequency response as the harmonic grounding impedance provided by HEM. Thus, it includes reactive and electromagnetic wave propagation effects.

5. Test Cases

This section presents the test cases related to the comparison between two different grounding modeling, i.e., the low-frequency grounding resistance (R_{LF}) and the harmonic grounding impedance ($Z(\omega)$).

Let us consider a 1.2 km matched three-phase DS (Figure 1). The three-phase conductors' heights are 10, 11 and 12 m, respectively, while the shield wire height is 14 m. The horizontal distance between each conductor and the shield wire is 2.4 m. The conductors' diameter is 1.83 cm, while the shield wire diameter is 0.72 cm.

The span between each tower is 50 m. To consider the influence of the adjacent towers, a total of five towers are modeled in detail, because the towers in longer distance have little impact on the overvoltage. According to [52], for lightning-related phenomena adjacent towers place a moderate influence. Each tower is 14 m high and with a base diameter of 0.5 m. According to Equation (3), a value of $Z_c = 244.17 \ \Omega$ is considered. The propagation velocity along the tower is considered to be $0.8c$ [53,54]. Moreover, the insulators are modeled taking into account their parasitic capacitance. The value of parasitic capacitance of each insulator is 7.68 nF, and they were calculated considering the information in [55–57].

Each tower is grounded with a grounding system as shown in Figure 2. This is a typical configuration for grounding distribution networks in the State of Minas Gerais, Brazil. It consists of three vertical rods 2.5 m long interconnected by a horizontal galvanized steel cable 6 m long. The vertical rods are copper-plated steel, with a diameter of 15 mm.

The equivalent circuit of the system composed of a three-phase distribution line, tower and grounding system is shown in Figure 3.

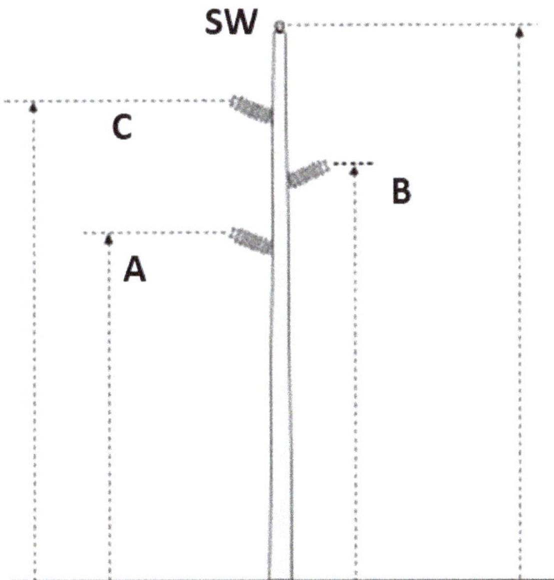

Figure 1. Line configuration.

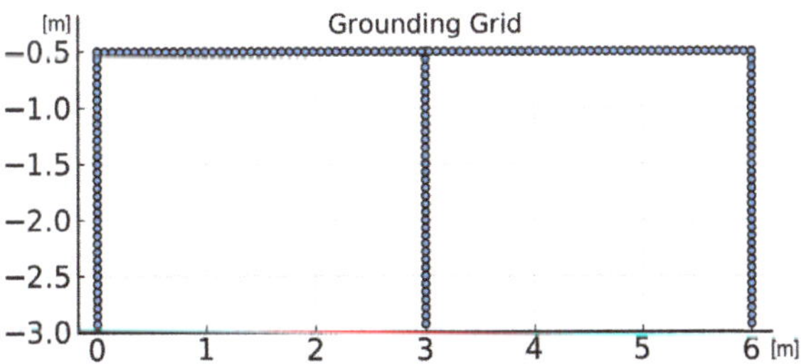

Figure 2. Grounding grid of the distribution tower.

Two cases for the soil parameters will be considered. The soil conductivity and permittivity will be frequency-dependent according to [31], where $\sigma_0 = 10$ mS/m and 1 mS/m, respectively. These cases correspond to two different grounding harmonic responses according to the grounding modeling proposed in Section 4. Figures 4 and 5 show $Z(\omega)$ and R_{LF} of the two considered cases. Based on the behaviors described in these figures, it is possible to verify that: (i) grounding can only be represented by R_{LF} in the low-frequency range, where $Z(\omega)$ tends to R_{LF}; (ii) the limit frequency of the low-frequency range increases with a reduction in conductivity; (iii) in the intermediate-frequency range there is a predominance of capacitive behavior of the grounding, verified by the decrease of $Z(\omega)$ in relation to the R_{LF}; (iv) the limit frequency of the intermediate frequency range also increases with the reduction in conductivity and (v) only in the high-frequency range inductive effect is predominant, mainly for higher conductivity values. Thus, the response of the system under study (DS and grounding) will be a direct function of the frequency spectrum of the electromagnetic signal that requests it. As a consequence, it is expected that the overvoltages in the insulator string are sensitive to grounding modeling.

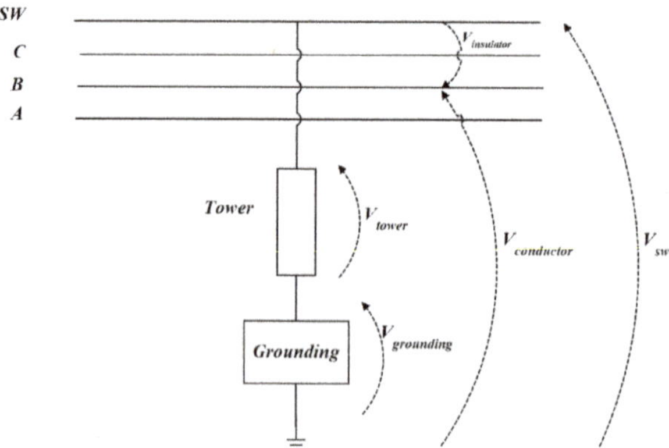

Figure 3. Equivalent circuit of the power system, tower and grouding.

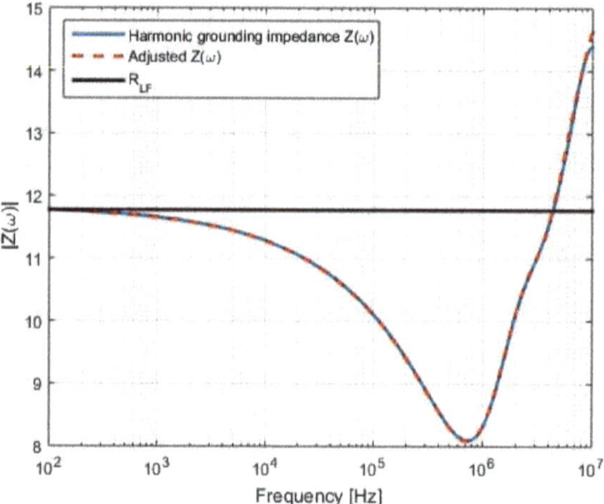

Figure 4. Grounding harmonic impedance $Z(\omega)$ with $\sigma = 10$ mS/m. Adjusted Z refers to the impedance obtained with the equivalent circuit.

When we consider a grounding model described by R_{LF}, the implementation in the EMT-type software is trivial, while when we consider the harmonic grounding impedance, it is possible to obtain the synthesis of the electric circuit to be implemented in the EMT-type software by using the approach presented in Section 4.

The general layout of the circuit obtained from the Vector fitting approach is described in Figure 6, while the values of the passive circuit are proposed in Tables 1 and 2 for $\sigma = 10$ mS/m and in Tables 3 and 4 for $\sigma = 1$ mS/m. It is worth mentioning that these equivalent circuits are mathematical models that have a frequency response very close to $Z(\omega)$, but their electrical parameters do not have physical consistency, it is also important mentioning that although some elements may have negative values, the circuit is passive in overall. Hence the existence of negative values for resistance, inductance and capacitance in Tables 1–4 do not mean that the circuit is not passive. The i-index appearing in Tables 1–4 refers to the electrical branch.

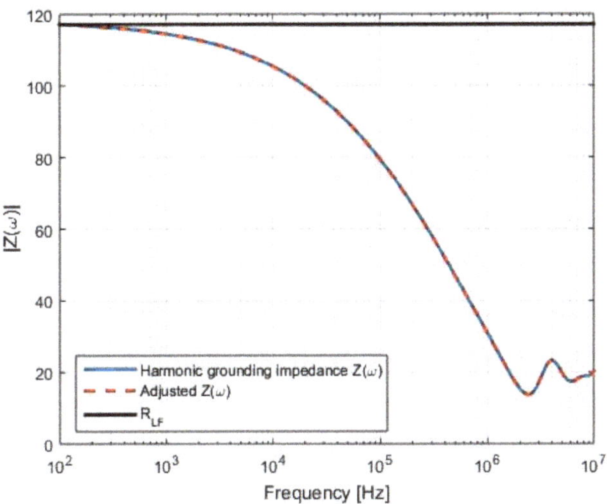

Figure 5. Grounding harmonic impedance $Z(\omega)$ with $\sigma = 1$ mS/m. Adjusted Z refers to the impedance obtained with the equivalent circuit.

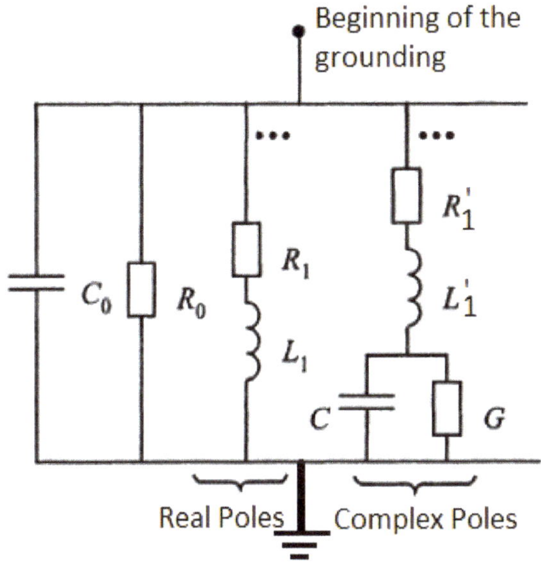

Figure 6. General layout of the grounding circuit for the harmonic grounding impedance.

Table 1. Elements of Real Poles of the grounding circuit. $\sigma = 10$ mS/m.

i	Resistance [Ω]	Inductance [mH]	Capacitance [μF]
0	3.33×10^{15}	-	3.00×10^{-10}
1	-4.23×10^3	-1.44×10^4	-
2	-3.12×10^3	-2.25×10^3	-
3	-1.83×10^3	-3.96×10^2	-
4	-1.03×10^3	-70.87	-
5	-5.81×10^2	-12.83	-
6	-3.23×10^2	-2.33	-
7	-1.75×10^2	-0.41	-
8	-83.82	-6.38×10^2	-
9	-5.40×10^3	-9.35×10^{-4}	-
10	11.77	3.09×10^{-4}	-
11	-7.65	-5.21×10^{-5}	-
12	6.21	1.32×10^{-5}	-

Table 2. Elements of Complex Poles of the grounding circuit. $\sigma = 10$ mS/m.

i	Resistance [Ω]	Inductance [mH]	Capacitance [μF]	Conductance [mS]
1	-2.35	1.25×10^{-3}	6.70×10^{-3}	1.26×10^2

Table 3. Elements of Real Poles of the grounding circuit. $\sigma = 1$ mS/m.

i	Resistance [Ω]	Inductance [mH]	Capacitance [μF]
0	3.33×10^{15}	-	3.00×10^{-10}
1	-6.35×10^3	-5.67×10^3	-
2	-8.44×10^3	-1.19×10^3	-
3	-1.30×10^3	-3.66×10^1	-
4	-2.75×10^3	-1.74×10^1	-
5	-2.78×10^2	-4.23×10^{-1}	-
6	-8.96×10^2	-3.39×10^{-1}	-
7	-3.13×10^1	-1.98×10^{-3}	-
8	1.77×10^{-5}	5.00×10^{-5}	-

Table 4. Elements of Complex Poles of the grounding circuit. $\sigma = 1$ mS/m.

i	Resistance [Ω]	Inductance [mH]	Capacitance [μF]	Conductance [mS]
1	1.58×10^1	2.28×10^{-3}	1.89×10^{-3}	5.18
2	2.28×10^3	1.11×10^{-2}	2.34×10^{-6}	-0.42
3	3.35×10^2	-9.12×10^{-3}	-2.16×10^{-5}	-1.12

To compare the grounding modeling, 12 different tests have been implemented (Table 5), each one differing for the soil conductivity, stroke location and stroke type (first or subsequent). The stroke location is always placed in front of the middle of the line, Figure 7 illustrates the distance between the lightning-channel and the tower under study. It is important to highlight that if the closest point of the DS to the stroke location is in the mid-span, it would, naturally, change the maximum overvoltage but not the perceptual differences between the approaches. The lightning return stroke channel is characterized by a height of 8 km and a speed equal to one-half the speed of light in a vacuum. The channel-base current is modeled as a sum of two Heidler's functions as in Equation (6), with parameters reported in Table 6. The representation of the channel-base current is proposed in Figures 8 and 9.

$$I_0(t) = \frac{I_{01}}{\eta_1} \frac{\left(\frac{t}{\tau_{11}}\right)^{n_1}}{1 + \left(\frac{t}{\tau_{11}}\right)^{n_1}} e^{-\frac{t}{\tau_{12}}} + \frac{I_{02}}{\eta_2} \frac{\left(\frac{t}{\tau_{21}}\right)^{n_2}}{1 + \left(\frac{t}{\tau_{21}}\right)^{n_2}} e^{-\frac{t}{\tau_{22}}} \tag{6}$$

being

$$\eta_i = \exp\left(-\frac{\tau_{i1}}{\tau_{i2}}\left(n_i \frac{\tau_{i2}}{\tau_{i1}}\right)^{\frac{1}{n_i}}\right) \tag{7}$$

Table 5. Test details.

Test	σ [S/m]	Stroke Distance [m]	Stroke Type
T1	0.01	60	First
T2	0.01	200	First
T3	0.01	2000	First
T4	0.001	60	First
T5	0.001	200	First
T6	0.001	2000	First
T7	0.01	60	Subsequent
T8	0.01	200	Subsequent
T9	0.01	2000	Subsequent
T10	0.001	60	Subsequent
T11	0.001	200	Subsequent
T12	0.001	2000	Subsequent

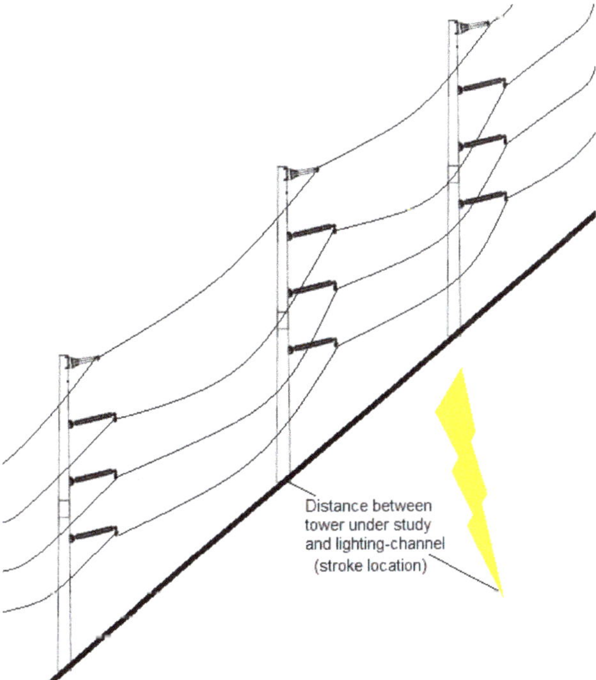

Figure 7. Sketch that illustrates the distance between the tower under study and the lightning stroke Location.

Table 6. Heidler's current parameters.

Parameter	First	Subsequent
I_{01} [kA]	28.0	10.7
τ_{11} [µs]	1.8	0.22
τ_{12} [µs]	95.0	2.5
n_1	2	2
I_{02} [kA]	–	6.5
τ_{21} [µs]	–	2.1
τ_{22} [µs]	–	230.0
n_2	–	2

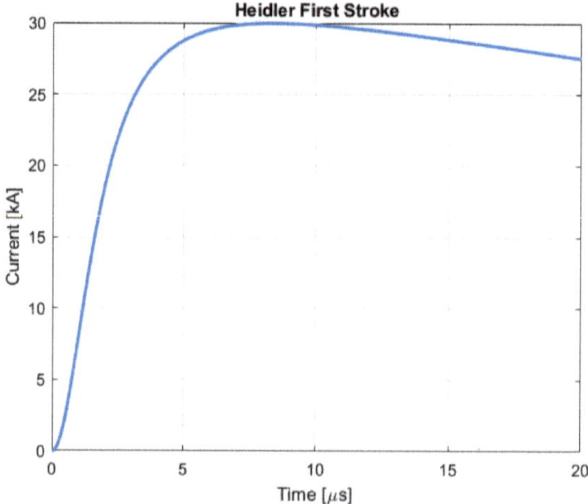

Figure 8. Channel-base current: Heidler's current-first stroke.

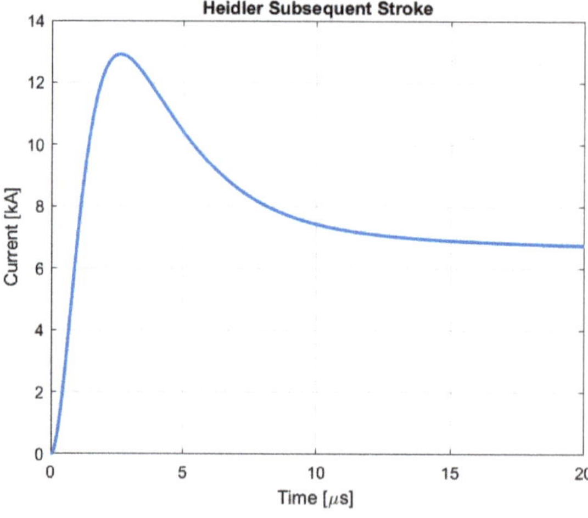

Figure 9. Channel-base current: Heidler's current-subsequent stroke.

6. Results

In this section, the results for the test cases of Table 5 are presented, showing the voltage across the phase B insulator string ($V_{insulator}$ in Figure 3) and the voltage difference occurring on the grounding system ($V_{grounding}$ in Figure 3).

Figures 10–15 show the results for tests T1–T6, corresponding to a typical first stroke. The main differences in terms of voltage across the insulator can be observed considering a low soil conductivity (Figures 13–15) and near stroke locations (60 m). This is extremely important because the closer the stroke location, the higher (and the more dangerous) the induced voltage. For example, let us consider Test T4 (Figure 13). If we use the low-frequency grounding resistance (R_{LF}) as grounding model, the maximum induced voltage across the insulator string is 115.12 kV, while if we consider the harmonic grounding impedance ($Z(\omega)$), which represents in a better way the reality, the voltage is 119.40 kV. This shows how the difference in the modeling could lead to either a fault or not across the insulator strings.

On the other hand, when the harmonic grounding impedance model presents a voltage across the insulator higher with respect to the R_{LF} case, the voltage on the grounding system is lower. This can be explained as follows: let us consider Figure 3; the voltage difference occurring on the insulator string is

$$V_{insulator} = V_{conductor} - V_{sw} \tag{8}$$

It is reasonable to assume that the voltage on the conductor does not change in a meaningful way. Considering the two different approaches (based on grounding system modeling), the only difference is the current flowing in the shield wire conductor causing a different coupling with the phase conductor. Even if not negligible, the coupling between conductors does not represent the dominant aspect in the lightning-induced voltages (which is the electric field illuminating the conductor). Consequently, $V_{insulator} + V_{sw}$ is almost constant. The shield wire voltage is:

$$V_{sw} = V_{tower} + V_{grounding} \tag{9}$$

with the same current, V_{tower} is constant in the two cases but $V_{grounding}$ varies because the impedance varies according to Figures 4 and 5 for $\sigma = 10$ mS/m and 1 mS/m, respectively. Let us consider the most critical case, i.e., $\sigma = 1$ mS/m: from Figure 5 it is clear that for each considered frequency $Z(\omega) < R_{LF}$, thus with the same current the voltage on the grounding system is lower if we consider the harmonic impedance $Z(\omega)$ and consequently also V_{sw} is lower. Since $V_{insulator} + V_{sw} = constant$, if V_{sw} decreases , $V_{insulator}$ increases. This aspect is confirmed in Tests T4-T5-T6, T10-T11-T12.

The results for subsequent strokes can be observed in Figures 16–21. The results are in agreement with the previous ones, confirming a significant increase of the maximum voltage if the equivalent circuit ($Z(\omega)$) is taken into account, especially if the soil conductivity is low. Moreover, the percentage increase considering the harmonic grounding impedance with respect to the low-frequency resistance is much more significant with respect to the first stroke case considering $\sigma = 1$ mS/m. It is expected since the subsequent strokes are faster. Thus, it has a higher frequency spectrum (the region where there are the highest differences between R_{LF} and $Z(\omega)$). Additionally, it is important to highlight that after a while, both overvoltages, considering R_{LF} and $Z(\omega)$, tend to the same value (first and subsequent strokes). For instance, if we consider the T10 and use the R_{LF} as grounding model, the maximum induced voltage across the insulator string is 66 kV, while if we consider the $Z(\omega)$, the voltage is 81 kV. This shows how the modeling difference could lead to either a fault or not across the insulator strings, especially for subsequent strokes.

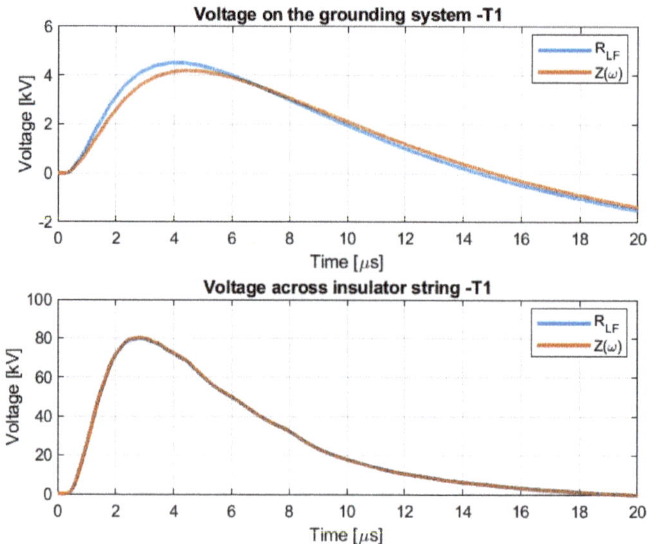

Figure 10. Test T1—Voltage on the grounding system and on the insulator of phase B. Comparison between the two models.

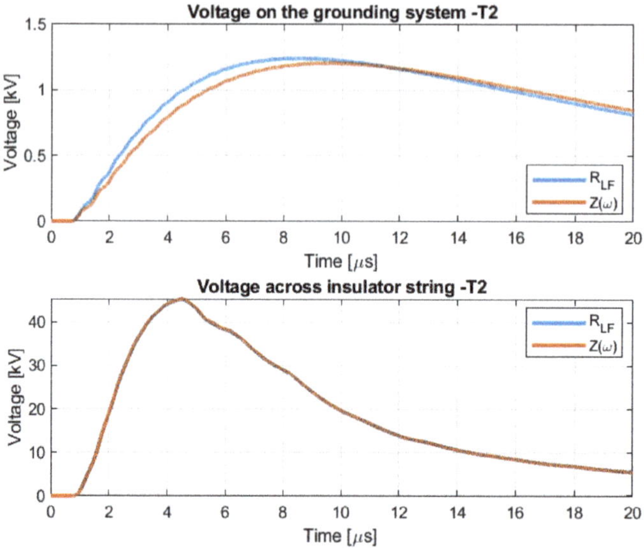

Figure 11. Test T2—Voltage on the grounding system and on the insulator of phase B. Comparison between the two models.

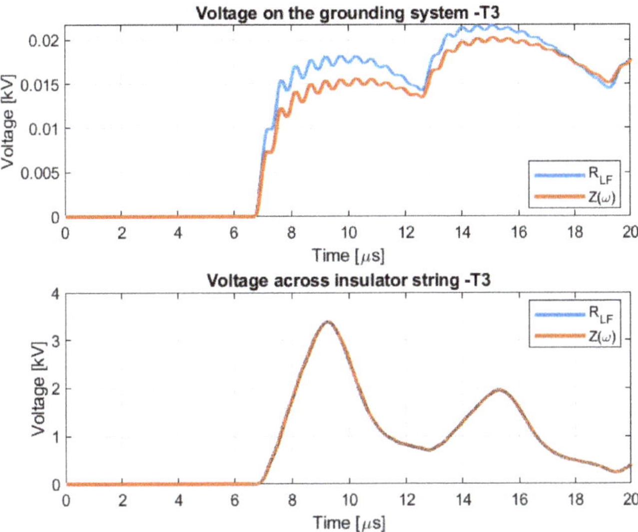

Figure 12. Test T3—Voltage on the grounding system and on the insulator of phase B. Comparison between the two models.

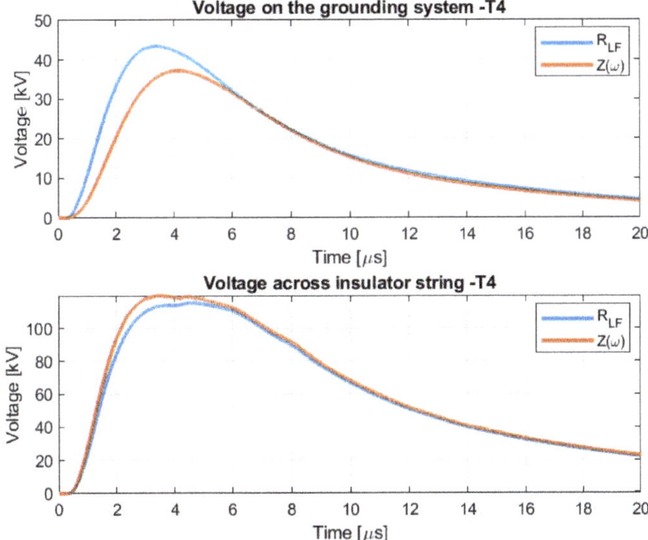

Figure 13. Test T4—Voltage on the grounding system and on the insulator of phase B. Comparison between the two models.

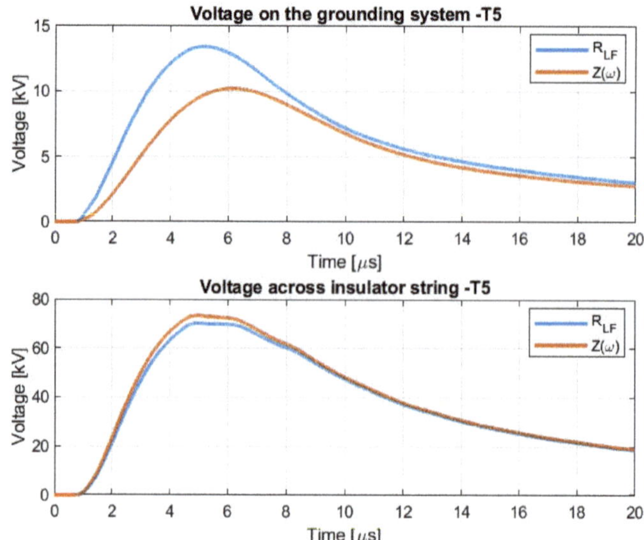

Figure 14. Test T5—Voltage on the grounding system and on the insulator of phase B. Comparison between the two models.

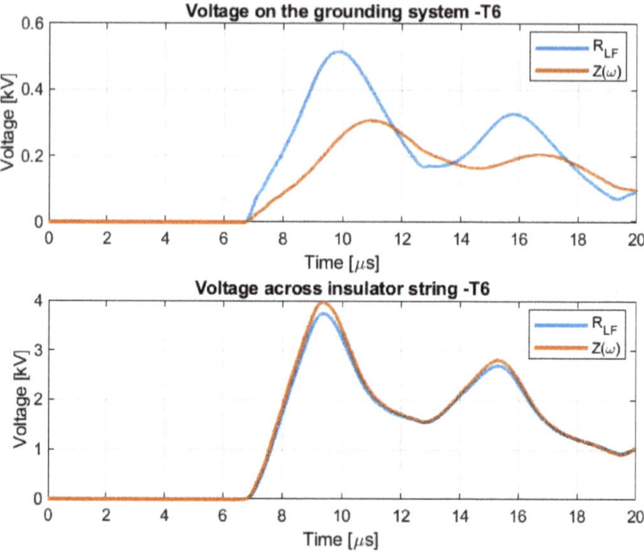

Figure 15. Test T6—Voltage on the grounding system and on the insulator of phase B. Comparison between the two models.

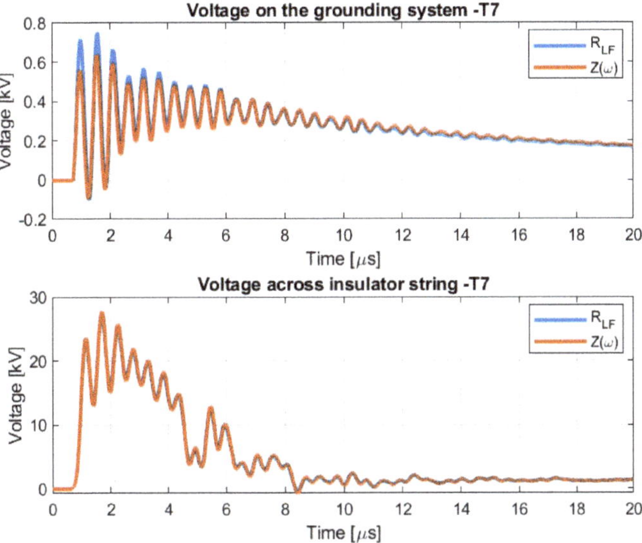

Figure 16. Test T7—Voltage on the grounding system and on the insulator of phase B. Comparison between the two models.

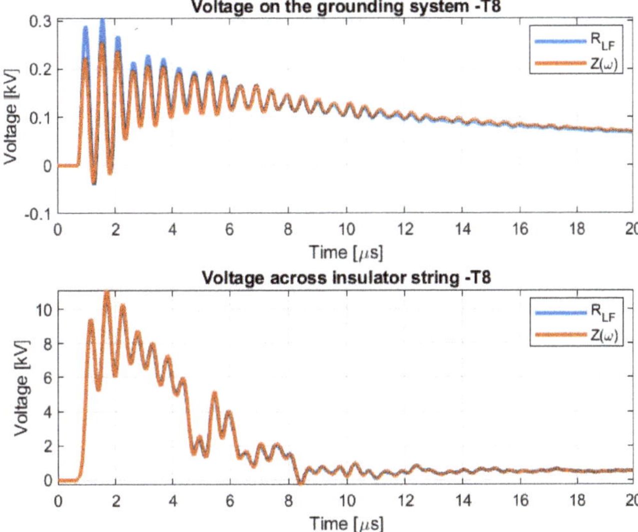

Figure 17. Test T8—Voltage on the grounding system and on the insulator of phase B. Comparison between the two models.

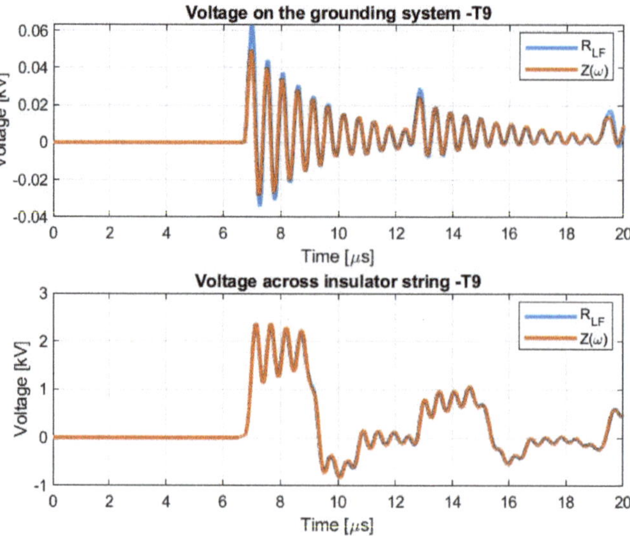

Figure 18. Test T9—Voltage on the grounding system and on the insulator of phase B. Comparison between the two models.

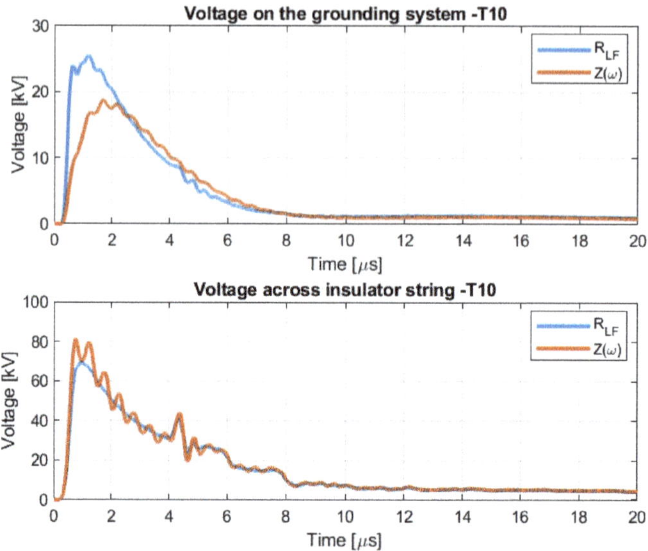

Figure 19. Test T10—Voltage on the grounding system and on the insulator of phase B. Comparison between the two models.

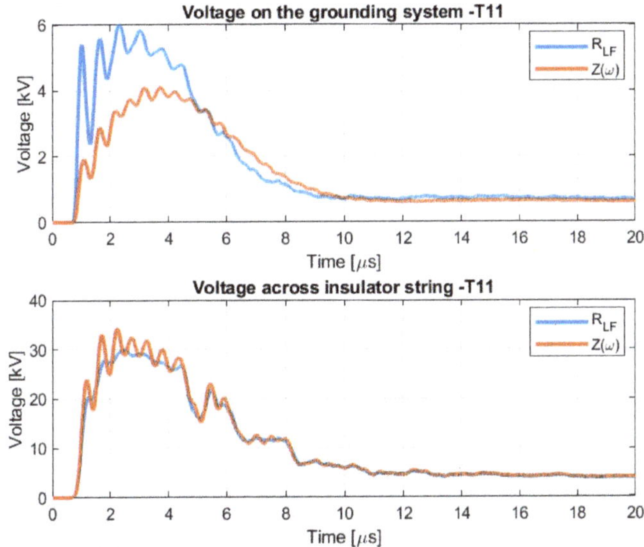

Figure 20. Test T11—Voltage on the grounding system and on the insulator of phase B. Comparison between the two models.

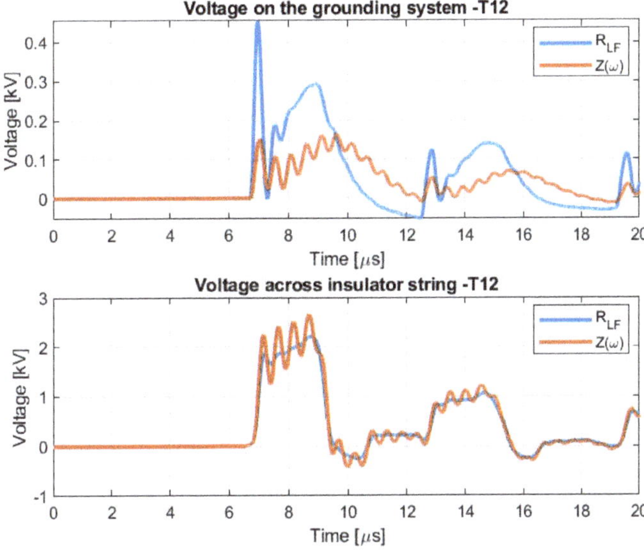

Figure 21. Test T12—Voltage on the grounding system and on the insulator of phase B. Comparison between the two models.

Finally, Table 7 shows the percentage increase in the maximum voltage across the phase B insulator considering the harmonic grounding impedance ($Z(\omega)$) with respect to the low-frequency grounding resistance (R_{LF}). According to the previous considerations, the differences are almost negligible if the soil conductivity is high (tests T1–T3 and T7–T9), but they become consistent when the soil conductivity decreases (tests T4–T6 and T10–T12). This behavior is more evident for close stroke location (test T4 and T10).

Table 7. Maximum voltage across the insulator. Percentage increase considering the harmonic grounding impedance ($Z(\omega)$) with respect to the low-frequency grounding resistance (R_{LF}).

Test	Voltage Insulator Increase [%]
T1	0.56
T2	0.26
T3	0.08
T4	3.50
T5	4.95
T6	6.76
T7	1.73
T8	1.80
T9	1.26
T10	22.54
T11	11.31
T12	23.06

7. Conclusions

Lightning induced-voltages are usually computed considering only the low-frequency grounding resistance when one considers the grounding system of the distribution tower. This work presented the impact of two different models for the grounding system of distribution line towers on the lightning-induced voltage on the phase insulators computation. The comparison between the low-frequency grounding resistance (R_{LF}) and the equivalent circuit corresponding to the wideband grounding frequency response ($Z(\omega)$) shows that considering only R_{LF} may lead to non-negligible underestimation of the maximum induced voltage. This aspect is more evident for subsequent strokes in the case of close stroke locations and low soil conductivities, which represents, by the way, one of the configurations when the lightning-induced voltages on a distribution line are high and potentially dangerous. On the other hand, for high soil conductivity, the differences between the two models are negligible. Future work will extend this analysis to the evaluation of a distribution line lightning performance to check whether this trend is also confirmed when dealing with statistical calculations. Additionally, in future works it is expected to include the coupling between the lightning channel and the grounding electrodes, similarly, as presented in [58].

Author Contributions: Conceptualization, D.M. and R.A.R.d.M.; methodology, D.M.; software, R.P. and R.A.R.d.M.; validation, D.M. and M.A.d.O.S.; formal analysis, D.M. and M.A.d.O.S.; writing—original draft preparation, D.M. and R.A.R.d.M.; writing—review and editing, R.P. and M.A.d.O.S.; All authors have read and agreed to the published version of the manuscript.

Funding: This research received no external funding.

Institutional Review Board Statement: Not applicable.

Informed Consent Statement: Not applicable.

Conflicts of Interest: The authors declare no conflict of interest.

References

1. Cooray, V.; Scuka, V. Lightning-induced overvoltages in power lines: Validity of various approximations made in overvoltage calculations. *IEEE Trans. Electr. Compatib.* **1998**, *40*, 355–363. [CrossRef]
2. Borghetti, A.; Nucci, C.A.; Paolone, M. An Improved Procedure for the Assessment of Overhead Line Indirect Lightning Performance and Its Comparison with the IEEE Std. 1410 Method. *IEEE Trans. Power Deliv.* **2007**, *22*, 684–692. [CrossRef]
3. Ren, H.; Zhou, B.; Rakov, V.A.; Shi, L.; Gao, C.; Yang, J. Analysis of Lightning-Induced Voltages on Overhead Lines Using a 2-D FDTD Method and Agrawal Coupling Model. *IEEE Trans. Electr. Compatib.* **2008**, *50*, 651–659. [CrossRef]
4. Paulino, J.O.S.; Barbosa, C.F.; Lopes, I.J.S.; Boaventura, W.d.C. An Approximate Formula for the Peak Value of Lightning-Induced Voltages in Overhead Lines. *IEEE Trans. Power Deliv.* **2010**, *25*, 843–851. [CrossRef]

5. Paolone, M.; Rachidi-Haeri, F.; Nucci, C.A. *IEEE Guide for Improving the Lightning Performance of Electric Power Overhead Distribution Lines*; IEEE Std 1410-2010 (Revision of IEEE Std 1410-2004); IEEE: New York, NY, USA, 2011; pp. 1–73. [CrossRef]

6. Silveira, F.H.; De Conti, A.; Visacro, S. Voltages Induced in Single-Phase Overhead Lines by First and Subsequent Negative Lightning Strokes: Influence of the Periodically Grounded Neutral Conductor and the Ground Resistivity. *IEEE Trans. Electr. Compatib.* **2011**, *53*, 414–420. [CrossRef]

7. Paulino, J.O.S.; Barbosa, C.F.; Lopes, I.J.S.; Boaventura, W.d.C. The Peak Value of Lightning-Induced Voltages in Overhead Lines Considering the Ground Resistivity and Typical Return Stroke Parameters. *IEEE Trans. Power Deliv.* **2011**, *26*, 920–927. [CrossRef]

8. Andreotti, A.; Pierno, A.; Rakov, V.A. An Analytical Approach to Calculation of Lightning Induced Voltages on Overhead Lines in Case of Lossy Ground—Part II: Comparison With Other Models. *IEEE Trans. Power Deliv.* **2013**, *28*, 1224–1230. [CrossRef]

9. Paulino, J.; Barbosa, C.; Lopes, I.; Boaventura, W. Assessment and analysis of indirect lightning performance of overhead lines. *Electr. Power Syst. Res.* **2015**, *118*, 55–61. [CrossRef]

10. Borghetti, A.; Napolitano, F.; Nucci, C.A.; Tossani, F. Influence of the return stroke current waveform on the lightning performance of distribution lines. *IEEE Power Energy Soc. Gener. Meet.* **2017**, *1*. [CrossRef]

11. Piantini, A. Extension of the Rusck Model for Calculating Lightning-Induced Voltages on Overhead Lines Considering the Soil Electrical Parameters. *IEEE Trans. Electr. Compatib.* **2017**, *59*, 154–162. [CrossRef]

12. Piantini, A. Analysis of the effectiveness of shield wires in mitigating lightning-induced voltages on power distribution lines. *Electr. Power Syst. Res.* **2018**, *159*, 9–16. [CrossRef]

13. Paulino, J.O.S.; Barbosa, C.F. Effect of high-resistivity ground on the lightning performance of overhead lines. *Electr. Power Syst. Res.* **2019**, *172*, 253–259. [CrossRef]

14. Paulino, J.O.S.; Barbosa, C.F. On Lightning-Induced Voltages in Overhead Lines Over High-Resistivity Ground. *IEEE Trans. Electr. Compatib.* **2019**, *61*, 1499–1506. [CrossRef]

15. Andreotti, A.; Araneo, R.; Mahmood, F.; Piantini, A.; Rubinstein, M. An Analytical Approach to Assess the Influence of Shield Wires in Improving the Lightning Performance due to Indirect Strokes. *IEEE Trans. Power Deliv.* **2020**, *1*. [CrossRef]

16. Brignone, M.; Mestriner, D.; Procopio, R.; Piantini, A.; Rachidi, F. Evaluation of the Mitigation Effect of the Shield Wires on Lightning Induced Overvoltages in MV Distribution Systems Using Statistical Analysis. *IEEE Trans. Electr. Compatib.* **2018**, *60*, 1400–1408. [CrossRef]

17. Schroeder, M.A.O.; de Barros, M.T.C.; Lima, A.C.; Afonso, M.M.; Moura, R.A. Evaluation of the impact of different frequency dependent soil models on lightning overvoltages. *Electr. Power Syst. Res.* **2018**, *159*, 40–49. [CrossRef]

18. Visacro, S.; Soares, A.; Schroeder, M. An interactive computational code for simulation of transient behavior of electric system components for lightning currents. In Proceedings of the 26th International Conference Lightning Protection, Uppsala, Sweden, 23–26 June 2008; pp. 732–737.

19. Visacro, S.; Soares, A. HEM: A model for simulation of lightning-related engineering problems. *IEEE Trans. Power Deliv.* **2005**, *20*, 1206–1208. [CrossRef]

20. Gustavsen, B.; Semlyen, A. Rational approximation of frequency domain responses by vector fitting. *IEEE Trans. Power Deliv.* **1999**, *14*, 1052–1061. [CrossRef]

21. Gustavsen, B. Fast Passivity Enforcement for Pole-Residue Models by Perturbation of Residue Matrix Eigenvalues. *IEEE Trans. Power Deliv.* **2008**, *23*, 2278–2285. [CrossRef]

22. Brignone, M.; Delfino, F.; Procopio, R.; Rossi, M.; Rachidi, F. Evaluation of Power System Lightning Performance, Part I: Model and Numerical Solution Using the PSCAD-EMTDC Platform. *IEEE Trans. Electr. Compatib.* **2017**, *59*, 137–145. [CrossRef]

23. Farina, L.; Mestriner, D.; Procopio, R.; Brignone, M.; Delfino, F. The Lightning Power Electromagnetic simulator for Transient Overvoltages (LIGHT-PESTO) code: An user-friendly interface with the Matlab-Simulink environment. *IEEE Lett. Electr. Compatib. Pract. Appl.* **2020**, *1*. [CrossRef]

24. Brignone, M.; Procopio, R.; Mestriner, D.; Rossi, M.; Delfino, F.; Rachidi, F.; Rubinstein, M. Analytical Expressions for Lightning Electromagnetic Fields with Arbitrary Channel-Base Current—Part I: Theory. *IEEE Trans. Electr. Compatib.* **2020**, 1–9. [CrossRef]

25. Mestriner, D.; Brignone, M.; Procopio, R.; Rossi, M.; Delfino, F.; Rachidi, F.; Rubinstein, M. Analytical Expressions for Lightning Electromagnetic Fields With Arbitrary Channel-Base Current. Part II: Validation and Computational Performance. *IEEE Trans. Electr. Compatib.* **2020**, 1–8. [CrossRef]

26. Agrawal, A.K.; Price, H.J.; Gurbaxani, S.H. Transient response of multiconductor transmission lines excited by a nonuniform electromagnetic field. *IEEE Trans. Electr. Compatib.* **1980**, *2*, 119–129. [CrossRef]

27. Rachidi, F.; Loyka, S.; Nucci, C.; Ianoz, M. A new expression for the ground transient resistance matrix elements of multiconductor overhead transmission lines. *Electr. Power Syst. Res.* **2003**, *65*, 41–46. [CrossRef]

28. Sargent, M.A.; Darveniza, M. Tower Surge Impedance. *IEEE Trans. Power Apparatus Syst.* **1969**, *PAS-88*, 680–687. [CrossRef]

29. Martinez-Velasco, J.A. *Power System Transients: Parameter Determination*, 1st ed.; CRC Press: Boca Raton, FL, USA, 2010.

30. Baba, Y.; Rakov, V.A. Electromagnetic computation methods for lightning surge studies with emphasis on the FDTD method. *Cigré Tech. Brochures 785* **2019**, WG C4.37, 1–192.

31. Alipio, R.; Visacro, S. Modeling the Frequency Dependence of Electrical Parameters of Soil. *IEEE Trans. Electr. Compatib.* **2014**, *56*, 1163–1171. [CrossRef]

32. Alipio, R.; Visacro, S. Impulse Efficiency of Grounding Electrodes: Effect of Frequency-Dependent Soil Parameters. *IEEE Trans. Power Deliv.* **2014**, *29*, 716–723. [CrossRef]

33. Visacro, S.; Alipio, R.; Pereira, C.; Guimarães, M.; Schroeder, M.A.O. Lightning Response of Grounding Grids: Simulated and Experimental Results. *IEEE Trans. Electr. Compatib.* **2015**, *57*, 121–127. [CrossRef]
34. Kuhar, A.; Arnautovski-Toševa, V.; Grčev, L. High frequency enhancement of the hybrid electromagnetic model by implementing complex images. *J. Electr. Eng. Inf. Technol.* **2017**, *2*, 79–87.
35. Kuhar, A.; Arnautovski-Toševa, V.; Ololoska-Gagoska, L.; Grčev, L.; Markovski, B. Influence of segmentation on the precision of circuit based methods. *J. Electr. Eng. Inf. Technol.* **2018**, *3*, 148.
36. De Oliveira Schroeder, M.A.; de Moura, R.A.R.; Machado, V.M. A Discussion on Practical Limits for Segmentation Procedures of Tower-Footing Grounding Modeling for Lightning Responses. *IEEE Trans. Electr. Compatib.* **2020**, *62*, 2520–2527. [CrossRef]
37. Harrington, R.F. *Field Computation by Moment Methods*; Wiley-IEEE Press: Melbourne, FL, USA, 1993.
38. Grcev, L.; Grceva, S. On HF circuit models of horizontal grounding electrodes. *IEEE Trans. Electr. Compatib.* **2009**, *51*, 873–875. [CrossRef]
39. Arnautovski-Toseva, V.; Grcev, L. On the Image Model of a Buried Horizontal Wire. *IEEE Trans. Electr. Compatib.* **2016**, *58*, 278–286. [CrossRef]
40. Smith-Rose, R.L. The electrical properties of soils for alternating currents at radio frequencies. *Proc. R. Soc.* **1933**, *140*, 359–377.
41. Scott, H.S. Dieletric Constant and Electrical Conductivity Measurements of Moist Rocks: A New Laboratory Method. *J. Geophys. Res.* **1967**, *1*, 5101–5115. [CrossRef]
42. Weir, W.B. Automatic measurement of complex dielectric constant and permeability at microwave frequencies. *Proc. IEEE* **1974**, *62*, 33–36. [CrossRef]
43. Hipp, J.E. Soil electromagnetic parameters as functions of frequency, soil density, and soil moisture. *Proc. IEEE* **1974**, *62*, 98–103. [CrossRef]
44. McKim, H.; Walsh, J.E.; Arion, D. *Review of Techniques for Measuring Soil Moisture in SITU*; United States Army, Corps of Engineers, Cold Regions Research and Engineering Laboratory: Hanover, NH, USA, 1980.
45. Messier, M. *Another Soil Conductivity Model*; Internal Rep.; JAYCOR: Santa Barbara, CA, USA, 1985.
46. Campbell, J.E. Dielectric properties and influence of conductivity in soils at one to fifty megahertz. *Soil Sci. Soc. Am. J.* **1990**, *54*, 332–341. [CrossRef]
47. Portela, C. Measurement and modeling of soil electromagnetic behavior. In Proceedings of the 1999 IEEE International Symposium on Electromagnetic Compatability, Symposium Record (Cat. No.99CH36261), Seattle, WA, USA, 2–6 August 1999; Volume 2, pp. 1004–1009.
48. Visacro, S.; Alipio, R. Frequency Dependence of Soil Parameters: Experimental Results, Predicting Formula and Influence on the Lightning Response of Grounding Electrodes. *IEEE Trans. Power Deliv.* **2012**, *27*, 927–935. [CrossRef]
49. Moura, R.A.R.; Schroeder, M.A.O.; Pereira, T.M.; Barros, M.T.C.; Alipio, R.S.; Lima, A.C.S. Analysis of frequency-dependence of soil resistivity: Emphasis at low frequencies. In Proceedings of the International Conference on Grounding, Lightning Physics and Effects, GROUND 2018 and 8th LPE, Pirenópolis, Brazil, 8–10 October 2018.
50. Delfino, F.; Procopio, R.; Rossi, M.; Rachidi, F. Influence of frequency-dependent soil electrical parameters on the evaluation of lightning electromagnetic fields in air and underground. *J. Geophys. Res. Atmos.* **2009**, *114*. [CrossRef]
51. Chisholm, W.; Visacro, S.; Alipio, R.; Griffiths, H.; Haddad, M.; He, J.; Montana, J.; Herrera-Murcia, J.; Sekioka, S.; Yamamoto, K.; et al. Impact of Soil-Parameter Frequency Dependence on the Response of Grounding Electrodes and on the Lightning Performance of Electrical Systems. 2020. Available online: https://www.researchgate.net/publication/341579253_Impact_of_soil-parameter_frequency_dependence_on_the_response_of_grounding_electrodes_and_on_the_lightning_performance_of_electrical_systems (accessed on 23 March 2021).
52. Soares, A.; Schroeder, M.A.O.; Visacro, S. Transient voltages in transmission lines caused by direct lightning strikes. *IEEE Trans. Power Deliv.* **2005**, *20*, 1447–1452. [CrossRef]
53. IEEE Working Group. *IEEE Guide for Improving the Lightning Performance of Transmission Lines*; IEEE Std 1243-1997; IEEE Working Group: New York, NY, USA, 1997; pp. 1–44. [CrossRef]
54. Chisholm, W.; Chow, Y.; Srivastava, K. Lightning Surge Response of Transmission Towers. *IEEE Trans. Power Apparatus Syst.* **1983**, *102*, 3232–3242. [CrossRef]
55. Dhalaan, S.M.A.; Elhirbawy, M.A. Simulation of voltage distribution calculation methods over a string of suspension insulators. In Proceedings of the 2003 IEEE PES Transmission and Distribution Conference and Exposition (IEEE Cat. No.03CH37495), Dallas, TX, USA, 7–12 September 2003; Volume 3, pp. 909–914. [CrossRef]
56. Tonmitr, N.; Tonmitr, K.; Kaneko, E. The Comparison of the String Suspension Porcelain (5-6-7 insulators)/string Efficiency in Case of Dry and Wetted water Contamination Condition. *Appl. Mech. Mater.* **2015**, *781*, 280–283. [CrossRef]
57. Tonmitr, N.; Tonmitr, K.; Kaneko, E. The effect of controlling stray and disc capacitance of ceramic string insulator in the case of clean and contaminated conditions. *Procedia Comput. Sci.* **2016**, *86*, 333–336. [CrossRef]
58. Nazari, M.; Moini, R.; Fortin, S.; Dawalibi, F.P.; Rachidi, F. Impact of Frequency-Dependent Soil Models on Grounding System Performance for Direct and Indirect Lightning Strikes. *IEEE Trans. Electr. Compatib.* **2021**, *63*, 134–144. [CrossRef]

Article

Corona Effect Influence on the Lightning Performance of Overhead Distribution Lines

Daniele Mestriner * and Massimo Brignone

Naval, ICT and Electrical Engineering Department (DITEN), University of Genoa, Via Opera Pia 11a, 16145 Genoa, Italy; massimo.brignone@unige.it
* Correspondence: daniele.mestriner@edu.unige.it; Tel.: +39-333-797-3889

Received: 25 June 2020; Accepted: 15 July 2020; Published: 17 July 2020

Abstract: Overhead distribution lines can be seriously damaged from lightning events because both direct and indirect events can cause flashovers along the line. The lightning performance of such power lines is usually computed neglecting the effect of corona discharge along the conductors: in particular, the corona discharge determined by the indirect lightning event is taken into account only by few researchers because it can have meaningful impacts only in few cases. However, when we deal with overhead distribution lines with high Critical Flashover value (CFO) and small diameters, the corona discharge caused by indirect events has to be taken into account. This paper shows the effects of corona discharge in the lightning performance computation of overhead distribution lines. The analysis will involve different configurations in terms of line diameter and air conditions, focusing on the negative effect of corona discharge in the number of dangerous events that determine line flashovers.

Keywords: corona discharge; lightning protection; electromagnetic pulse; lightning-induced voltages; numerical codes

1. Introduction

Lightning is one of the most important issues in terms of protection of transmission and distribution lines [1]. Their protection requires an accurate evaluation of the insulation coordination system [2] as well as a correct computation of the number of dangerous events striking the line per year. The latter parameter is usually computed through the lightning performance procedure—a high number of lightning events, each of them characterized by different parameters extracted from their own Probability Density Function (PDF), is generated and their effects on the power system are computed through a simplified method [1,3,4] or through a numerical code [5–9]. The number of events that exceeds a threshold value depending on the line Critical Flashover value (CFO) is considered dangerous. Each event can be classified as a direct or indirect stroke in accordance to the electrogeometrical criterion [10]. When we deal with transmission lines, characterized by high CFO, the majority of the dangerous events is represented by direct strokes, while dealing with overhead distribution lines, indirect strokes are the most affecting category.

The lightning performance procedure of overhead distribution lines has been deeply studied and optimized by several researchers and usually requires the use of a numerical code. Among them, Reference [7] proposes a procedure based on LIOV code which takes into account a triangular waveform for the channel-base current and the possibility to consider complex power systems as well as a finite ground conductivity. The authors of Reference [11] provide an application of the recursive stratified sampling technique in order to reduce the computational effort. In Reference [6], the authors propose a procedure which can be extended to whatever channel-base current based on the construction of an electromagnetic field database.

The corona effect, that is, the process that describes an electrical discharge by the ionization of the fluid that surrounds a conductor, usually, is not considered in the lightning performance of distribution lines (i.e., taking into account both direct and indirect events): typically due to its occurrence only for small conductor diameters and in case of direct events, which represents the less significant part. However, as pointed out in this work, there are some cases where the corona discharge occurs also for indirect events, leading to a meaningful impact in the lightning performance evaluation.

When one deals with direct events, the corona discharge affects the surge propagation more than the ground resistivity but reduces the voltages induced on the line [12]. On the other side, when one deals with indirect strokes, Reference [12], *"computation results showed a significant increase in the amplitude of the induced voltages in presence of corona"*. The result has been confirmed by Reference [13]. The implementation of corona discharge in the computation of lightning-induced voltages has been based on two different strategies: (i) The concept of dynamic capacitance has been proposed in Reference [12] that is, when the induced voltage overcomes a threshold, the corona discharge determines an increase of the per unit length capacitance involved in the Agrawal model [14]; consequently, the dynamic capacitance mimics the experimental q-v characteristic of the corona discharge [15]. (ii) In References [16,17] the 3D–FDTD code simulates the corona discharge giving different values to the conductivity of the cell where corona effect is located; the main difference with the previous approach is that here the q-v characteristic results as an output calculation from the numerical integrations, while in the previous one appears as an input of the problem.

However, both approaches lead to the same conclusions, that is, the enhancement of the induced voltage. Unfortunately, to the best of author's knowledge, both approaches limit their studies to the evaluation of the effect of a single event striking in the proximity of the line and a parametric analysis aimed at relating the most affecting parameters (line diameter, environmental conditions) to the induced voltage is missing.

The aim of this work is to evaluate the effects of corona discharge in the lightning performance computation of an overhead distribution line and to make some sensitivity analysis on the parameters that mainly affect the enhancement of the number of dangerous events striking the line. The implementation of corona discharge will be based on the procedure described in Reference [5] and validated in Reference [18,19] .

The paper is structured as follows: Section 2 recalls the concepts of the corona discharge, Section 3 describes the implementation of corona discharge in the procedure developed in Reference [5] and Section 4 focuses on the lightning performance computation. Later, the sensitivity analysis on the main parameters (line diameter and environmental conditions) affecting the enhancement of dangerous events due to corona is proposed in Section 5. Finally, in Section 6, some conclusions are drawn.

2. The Corona Effect

According to Reference [20], the corona discharge can occur when the electric field in air in the vicinity of object at high voltages or exposed to high electric fields may overhelm the critical electric field able to create electron avalanches in air. The corona discharge can be either a positive of negative discharge and, according to Reference [21], the corona occurs on a conductor when the electric field on its surface is higher than the critical field E_c.

$$E_c = m \times 2.594 \times 10^6 \left(1 + \frac{0.1269}{r^{0.4346}}\right), \tag{1}$$

where m is a surface state coefficient quantifying the irregularities of the cable and generally deduced from tests. This formula has been developed assuming 20 °C, a pressure of 760 mmHg and a humidity of the air equal to 11 g/m^3. Please note that r is the radius of the conductor and is expressed in cm. It is important to notice that according to Reference [21] the variation of humidity changes the critical field E_c. With respect to (1) an increase of the air humidity to 18 g/m^3 leads to an increase of the critical electric field of 2%. Figure 4 of Reference [21] showed the variation of the critical electric field as a function of the air humidity.

In Reference [22], recently, the authors provided a new expression for the critical electric field under variable atmospheric conditions, which is here proposed for sake of completeness.

$$E_c = 31.53 \left(1 + \frac{A}{K^a n_s^b r^c}\right), \tag{2}$$

where

$$K = \delta^{1.01} \left(1 + 0.08 \left(\frac{H}{11}^{0.72} - 1\right)\right), \tag{3}$$

being δ the relative air density, H the air humidity, n_s a coefficient depending on the number of strands in the outer layer of the conductor and A, a, b, c coefficients depending on the voltage polarity (Table 1 of Reference [22]).

In addition to this, the corona discharge occurs if a free electron is available at the instant when the electric field overcomes the critical value in (1). It means that there is a certain time lag between the application of the electric field and the time of creation of a free electron in the gas volume [20]: this time is known as the *statistical time lag* or *inception time delay*. The statistical time lag decreases when the applied electric field increases .

Once a free electron is found, the corona discharge occurs, but its sustainment is achieved only if the electric field in front of the streamers is not lower than 4 to 5 kV/cm (positive streamers) and 11 to 18 kV/cm (negative streamers) [20].

From a macroscopic point of view, the corona discharge on the surface of a conductor can be viewed as an increase of the capacitance of the conductor, while the inductance remains constant due to the low conductivity of the corona region. The capacitance can be estimated from the q-v curve, which can be obtained by experimental tests or by models in literature. An example is proposed in Figure 9 of Reference [12].

The lower line of Figure 9 of [12] represents the q-v curve measured during the increase of the voltage (i.e., when $dv/dt > 0$), while the upper line represents the q-v curve measured during the decrease of the applied voltage. In this second phase, the slope, that is, the capacitance, is constant and equal to the geometrical capacitance of the line because it is well-known that corona discharge occurs only when the applied voltage derivative is positive. The lower line presents an increase of the slope with applied voltages higher than 130 kV, which denotes the occurrence of the corona discharge

3. Implementation of Corona Discharge

This section shows the implementation of the corona effect in the procedure proposed in Reference [5] following the one in Reference [12].

The main idea of this approach is to consider a dynamic capacitance in the Agrawal model. As in Reference [12], the dynamic capacitance has been described through the following equation, which aims at reproducing the experimental q-v characteristics.

$$C_{dyn}(x,t) = \begin{cases} C_0 & \text{for } u(x,t) < u_{th}(x,t) \\ C_0 \frac{(k_1 + k_2(u(x,t) - u_{th}(x,t)))}{u_{th}(x,t)} & \text{for } u(x,t) \geq u_{th}(x,t) \text{ and } \frac{du(x,t)}{dt} > 0, \end{cases} \tag{4}$$

where $k_1 > 1$ is related to the sudden change of the capacitance when the voltage exceeds the corona threshold u_{th} and $k_2 > 0$ is related to the gradual increase of the capacitance when the voltage is rising above the threshold. C_0 is the geometrical capacitance of the line and $u(x,t)$ is the voltage on a generic conductor at time instant t and at a distance x from the beginning of the line. The voltage threshold u_{th} is related to C_0, to the conductor radius r and to the critical electric field in (1), according to

$$u_{th} = \frac{2\pi\varepsilon_0 r}{C_0} E_c.$$ (5)

Here, as usual, ε_0 is the electric permittivity in vacuum.

4. Corona Effect Influence on the Lightning Performance

This section shows the enhancement of the number of dangerous lightning events due to the corona effect.

Let us consider a single-phase overhead distribution line whose details are available in Table 1. In order to avoid reflections, the line extremities are matched. Moreover, let us suppose a ground conductivity of 1 mS/m and a ground permittivity of 10.

Table 1. Line details.

Length [m]	Height [m]	Conductor Diameter [mm]
1000	10	10

According to Reference [6] and supposing a lightning channel height of 8 km and a propagation velocity along the lightning channel of $c_0/2$ (being c_0 the light speed in vacuum), the lightning performance of the overhead distribution line is computed as follows (for further details check Figure 1):

1. A counter n is initialized to 0.
2. A large number of events, able to guarantee the convergence of the Monte-Carlo procedure (here $n_{tot} = 10{,}000$,   is generated. Each one is characterized by a stroke location extracted from a uniform distribution, peak current and front duration extracted from the log-normal distributions proposed in Reference [1]. Please note that the channel-base current is assumed to be the typical Heidler's first stroke waveform with variable front duration [23].
3. Each event is classified as direct or indirect one according to the EGM criterion [24]. In any case the maximum induced voltage on the line is computed through the procedure in Reference [5].
4. For each event, the maximum voltage is compared with the line CFO. If it is greater than 1.5 CFO [1], the counter n is increased by one.
5. Once all the considered events have been evaluated, the total number of dangerous events n is obtained and the number of flashovers per year per 100 km of line is computed according to References [1,7]

$$F = 200 \frac{n}{n_{tot}} GFD y_{max},$$ (6)

where GFD is the ground flash density expressed as number of flashes per square kilometer per year ad y_{max} is the maximum value of the y-coordinate where the events are extracted. According to Reference [1], y_{max} is a function of the CFO and it is computed through the extended Rusck's formula [25], choosing as lightning current the maximum value (I_{max}) obtainable from the probabilistic density function.

$$y_{max} = \frac{38.8 I_{max}\left(h + \frac{0.15}{\sqrt{\sigma_g}}\right)}{CFO},$$ (7)

where σ_g is the ground conductivity.

For what concerns the corona effect, in the following we assume $m = 0.8$, $k_1 = 1.2$ and $k_2 = 4.8$ [12]. Consequently, $E_c = 35.49$ kV/cm and $u_{th} = 150$ kV.

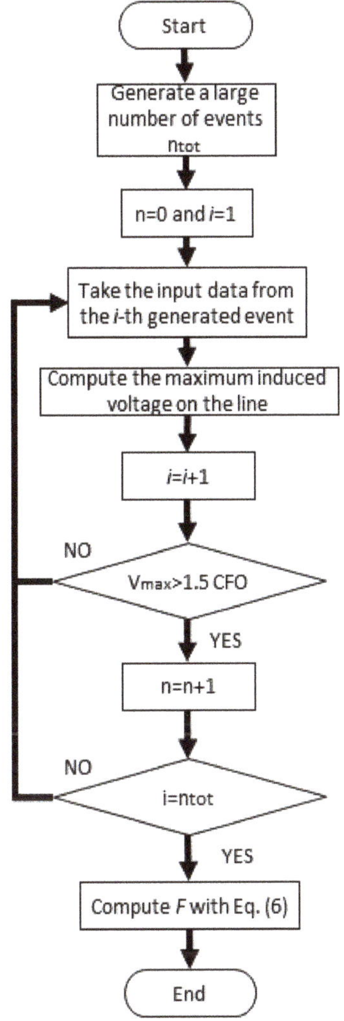

Figure 1. Flowchart of the lightning performance computation.

Figure 2 shows the number of dangerous events per 100 km of line per year with the corona effect (solid black line) and without (red dashed line). As can be easily seen, when we deal with overhead distribution lines characterizred by very low CFO ($\leq$100 kV), the corona discharge does not increase the number of flashovers. On the other side, when the CFO is higher, the influence of corona effect is meaningful. This result is in agreement with the conclusions of References [12,17], which have highlighted an increase of the lightning-induced voltages due to the corona effect. In particular, the enhancement can be caused by the decrease of the wave propagation velocity due to the increase of the line capacitance.

In order to quantify the importance of the corona effect, again in Figure 2 the lightning performance without introducing corona effect and considering a perfect conducting ground (cyano solid line) has been proposed. The percentage enhancement of the number of flashovers due to corona is comparable with the one caused by considering a lossy ground (Table 2) especially when the CFO is high—when the CFO overcomes 250 kV the dominant effect is ascribed to the corona discharge.

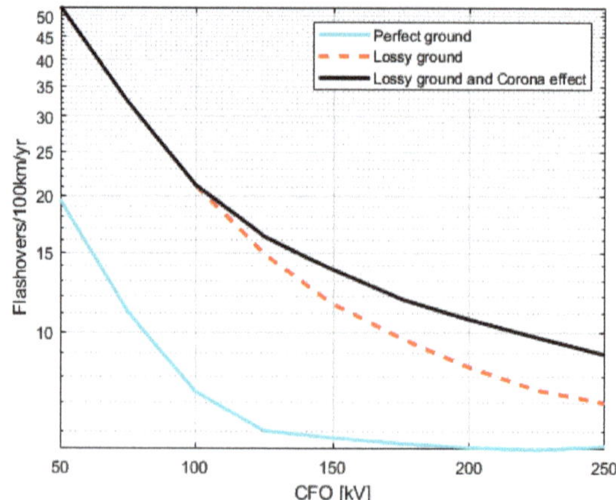

Figure 2. Number of dangerous events on an overhead distribution line. Comparison between perfect ground, lossy ground and lossy ground with corona discharge.

Table 2. Percentage enhancement—the second column describes how the lossy ground enhances the number of flashovers with respect to the PEC ground, while the third column describes how the corona effect enhances the number of flashovers with respect to the lossy ground case.

CFO [kV]	Enhancement Due to Lossy Ground [%]	Enhancement Due to Corona [%]
50	166.58	0
100	185.39	0
150	97.50	19.03
200	50.61	27.27
250	24.93	27.79

5. Sensitivity Analysis

This section aims at evaluating the flashovers variation due to corona discharge as a function of conductor surface state conditions, diameter and air humidity.

5.1. Surface Conditions

The effect of the surface conditions is taken into account by varying the surface state coefficient m. A minimum value of $m = 0.5$ and a maximum value of $m = 1$ are here considered [26]. Figure 3 shows the number of dangerous events per 100 km per year, computed for six different values for m in the selected range. For sake of completeness, the case treated in Figure 2 is here reported as the black line. The corresponding values of critical electric field and voltage can be found easily from Equations (1) and (5).

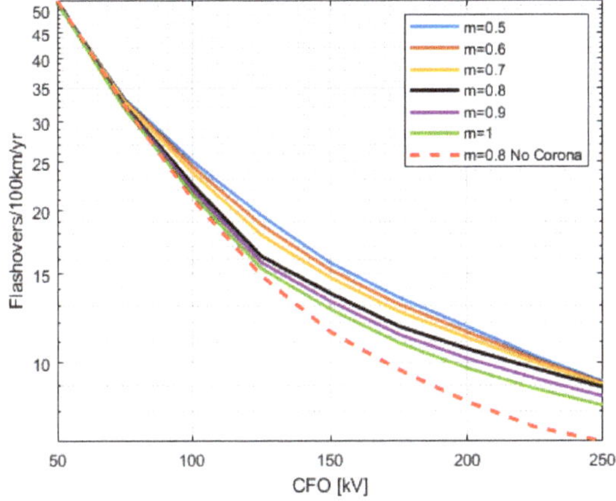

Figure 3. Comparison of the lightning performance of an overhead distribution line with different surface state conditions—the red dot line represents the case when the corona effect is not considered.

The number of dangerous events increase with the decrease of the surface coefficient since it corresponds to a linear decrease of the critical electric field.

Figure 4 shows the percentage variation (with respect to $m = 1$) of the number of flashovers as a function of the surface state coefficient for different line CFO.

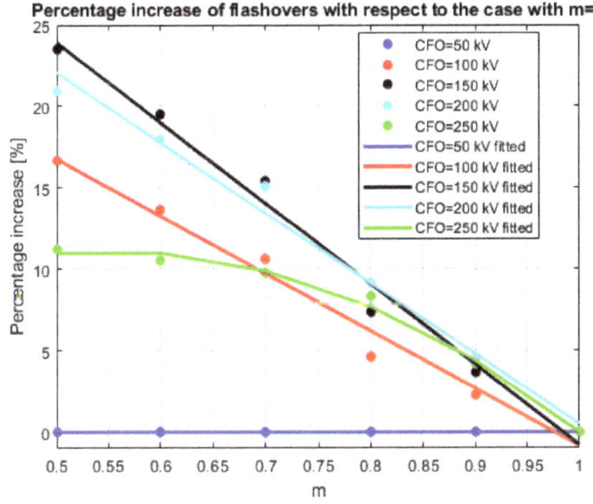

Figure 4. Percentage variation of the number of flashovers as a function of the surface state coefficient

Figure 4 shows that the percentage increase is negligible for $CFO = 50$ kV: this is obvious as the corona effect contributes to the enhancement of the lightning-induced voltages only when the travelling wave overcomes u_{th}, which is usually really higher than 50 kV. As a consequence, when $CFO = 50$ kV, those cases are already defined as dangerous whatever is the value of m. On the other hand, when $CFO = 250$ kV the curve can be well-described by a second-order polynomial because there is a sort of saturation for low values of m. This can be related to the fact that low values of m correspond to a decrease of u_{th}. In other words, the corona discharge contributes to the enhancement of the lightning-induced voltages also when the voltage on the line is low. However, the increase due to corona is not sufficient to overcome the threshold set with $CFO = 250$ kV, thus its effect on the enhancement of the flashovers number is less evident. Finally, the behaviour of the percentage increase for CFO $\in \{100, 150, 200\}$ kV is substantially linear, as shown by the fitting provided in Figure 4. This is in agreement with the linear variation of the critical electric field with respect to m. A decrease of the surface coefficient determines a linear decrease of u_{th}; the low value of u_{th} causes an higher occurrence of the corona discharge, causing an enhancement of the lightning-induced voltages and of the line flashovers. The higher increases can be noticed for $CFO = 150$ and 200 kV.

Equation (8) provides the general expression used for finding the fitting curve, that express the percentage increase as a function of m

$$P_{increase} = p_0 + p_1 m + p_2 m^2. \tag{8}$$

The values of the parameters p_0, p_1 amd p_2 are reported in Table 3, where also the R^2 index is shown, for quantify the reliability of the fitting.

Table 3. Fitting coefficients and R^2 index.

CFO	p_0	p_1	p_2	R^2
50	0.00	0.00	0.00	1.00
100	34.42	−35.24	0.00	0.98
150	48.65	−49.44	0.00	0.99
200	43.65	−43.13	0.00	0.99
250	−5.50	60.65	−55.09	0.99

5.2. Conductor Diameter

The effect of the conductor diameter is analysed, taking into account different values typical of overhead distribution lines (from 5 mm to 60 mm). The other line parameters have been described in the previous sections and in this framework we consider $m = 0.8$. Figure 5 shows that the influence of the corona discharge is negligible when the diameter is greater than 30 mm. Moreover, for very thin conductors ($d = 5$ mm), the corona discharge increases the number of flashovers even if the line CFO is very low (<100 kV): this can be ascribed to the low value of the critical electric field associated to such diameter.

Figure 6 analyses the percentage variation (with respect to the case with $d = 30$ mm) of the number of flashovers as a function of the conductor diameter for different line CFO.

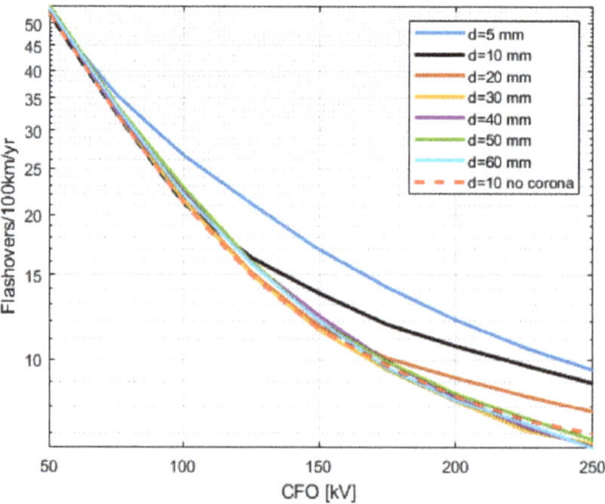

Figure 5. Comparison of the lightning performance of an overhead distribution line with different diameters—the red dot line represents the case when the corona effect is not considered.

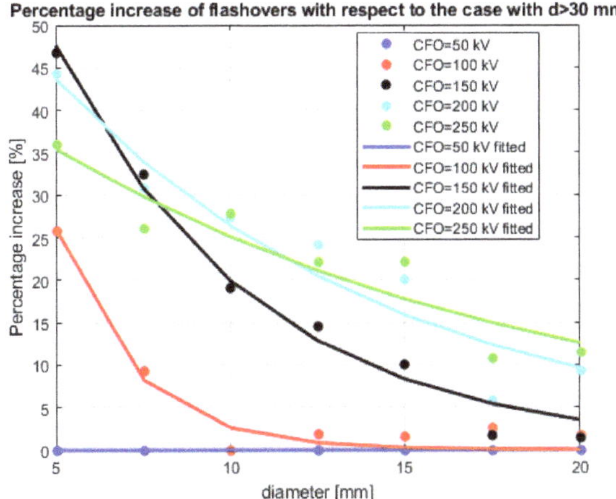

Figure 6. Percentage variation of the number of flashovers as a function of the conductor diameter.

Figure 6 confirms that, if the line CFO is 50 kV, the enhancement is negligible for the same reason presented in the previous subsection. Considering a line CFO of 100 kV and 150 kV corresponds to a consistent enhancement of the percentage for low values of the diameter (<10 mm) and a flatness of the increase if the diameter is greater than 10 mm. As a consequence the two curves can be well described by an exponential. This behaviour is related to the high values of u_{th} when we consider a diameter larger than 10 mm: in these cases $u_{th} \gg 100$ kV, thus every event that overcomes that threshold is already defined as dangerous. On the other hand, an increase of the line CFO (200, 250 kV) leads to a behaviour characterized by a low variation of its derivative.

Equation (9) provides the general expression used for finding the fitting curve that expresses the percentage increase as a function of d

$$P_{increase} = ae^{bd}. \tag{9}$$

The values of the parameters a and b are reported in Table 4, where also the R^2 index is shown, to quantify the reliability of the fitting.

The values of the R^2 indexes for CFO = 200, 250 kV are low as there are some deviations between the points and the curves of Figure 6. However, these deviations can be ascribed to the statistical procedure of the lightning performance and it is evident that the overall behaviour has a low variation in its derivative (low values of b).

Table 4. Fitting coefficients and R^2 index.

CFO	a	b	R^2
100	258.90	−0.46	0.95
150	114.00	−0.17	0.98
200	72.25	−0.1	0.92
250	49.95	−0.06	0.90

The sensitivity analysis on the diameter can help an user to evaluate, once that the CFO is set, how the number of flashovers due to corona increases with respect to diameter: for example, considering a line CFO of 100 kV, choosing a very thin diameter (5 mm) leads to a percentage increase of 25% but with a low cost for the conductor material. On the other hand increasing the diameter leads to higher costs but lower enhancement of dangerous events. It is important to notice that the cost is affected not only by the conductor diameter, but also by the ampacity and the rated tensile strength (RTS); as a consequence, also these two factors shall be taken into account while choosing the conductor.

As a final remark, attention shall be dedicated to the bundled conductors as their installation is frequent in overhead distribution lines. Although, their geometry can be seen in some ways as an equivalent conductor with a larger diameter, thus its capability of mitigating the corona discharge can be discussed as previously presented.

5.3. Air Humidity

The effect of the air humidity is here analyzed varying the critical electric field according to Reference [21]. A minimum value of 2 g/m^3 and a maximum value of 25 g/m^3 are here considered. Considering $m = 0.8$, a line diameter of 10 mm and the line details previously described, Figure 7 shows that the influence of the air humidity is substantially negligible. This is ascribed to the low variation of the critical electric field according to Reference [21].

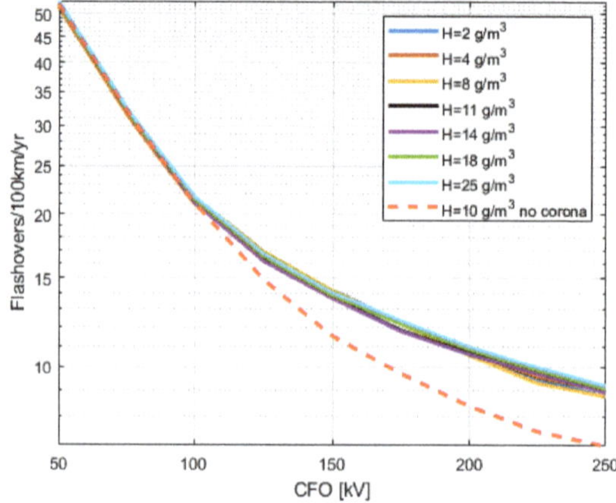

Figure 7. Percentage variation of the number of flashovers as a function of the air humidity—the red dot line represents the case when the corona effect is not considered.

6. Conclusions

This paper analysed the importance of corona discharge in the lightning performance computation, taking into account the effect of corona discharge caused by indirect events. The corona effect has been implemented in the FDTD code for lightning-induced voltages computation following the approach proposed in Reference [12]. The number of flashovers considering the corona discharge can be compared to the ones obtained considering the finite ground conductivity, which represents a consistent part of the total. Moreover, a sensitivity analysis on the main causes leading to the corona discharge and to the enhancement of the number of dangerous events has been proposed, involving the surface state coefficient, the conductor diameter and the air humidity. The surface state coefficient variation leads to a linear percentage increase of the number of flashovers, except for very CFO values. The line diameter has a meaningful effect when we consider values lower than 20 mm: the function that better represents the behaviour of the percentage increase with respect to the diameter is exponential for CFO < 100 kV and linear for high CFO values. Finally, the effect of the air humidity is negligible. The sensitivity analysis allows an user to forecast the overall behaviour of his line with respect to the considered variables and can be taken into considerations in future IEEE Guidelines or CIGRE Working Groups Technical Brochures.

Author Contributions: Conceptualization, D.M.; methodology, D.M. and M.B.; software, D.M. and M.B.; validation, D.M.; data curation, D.M.; writing—original draft preparation, D.M.; writing—review and editing, D.M. and M.B. All authors have read and agreed to the published version of the manuscript.

Funding: This research received no external funding.

Conflicts of Interest: The authors declare no conflict of interest.

Abbreviations

The following abbreviations are used in this manuscript:

CFO	Critical Flashover
EGM	Electro-Geometric Model
FDTD	Finite Difference Time Domain
GFD	Ground Flash Density
LIOV	Lightning-Induced OverVoltages
PDF	Probabilistic Density Function

References

1. *IEEE Guide for Improving the Lightning Performance of Electric Power Overhead Distribution Lines*; IEEE: Piscataway, NJ, USA, 2010; pp. 1–70.
2. Hileman, A. Insulation coordination for power systems. *IEEE Power Eng. Rev.* **1999**, *19*, 43. [CrossRef]
3. Rusck, S. *Induced Lightning Over-Voltages on Power-Transmission Lines with Special Reference to the Over-Voltage Protection of Low Voltage Networks*; Transactions of the Royal Intitute of Technology: Stockholm, Sweden, 1958.
4. Paulino, J.O.S.; Barbosa, C.F.; Lopes, I.J.S.; Boaventura, W.d.C. An Approximate Formula for the Peak Value of Lightning-Induced Voltages in Overhead Lines. *IEEE Trans. Power Deliv.* **2010**, *25*, 843–851. [CrossRef]
5. Brignone, M.; Delfino, F.; Procopio, R.; Rossi, M.; Rachidi, F. Evaluation of power system lightning performance, part I: Model and Numerical Solution Using the Pscad-Emtdc platform. *IEEE Trans. Electromagn. Compat.* **2016**, *59*, 137–145. [CrossRef]
6. Brignone, M.; Delfino, F.; Procopio, R.; Rossi, M.; Rachidi, F. Evaluation of power system lightning performance—Part II: Application to an overhead distribution network. *IEEE Trans. Electromagn. Compat.* **2016**, *59*, 146–153. [CrossRef]
7. Borghetti, A.; Nucci, C.A.; Paolone, M. An improved procedure for the assessment of overhead line indirect lightning performance and its comparison with the IEEE Std. 1410 method. *IEEE Trans. Power Deliv.* **2006**, *22*, 684–692. [CrossRef]
8. Nucci, C. The lightning induced over-voltage (LIOV) code. In Proceedings of the 2000 IEEE Power Engineering Society Winter Meeting. Conference Proceedings (Cat. No. 00CH37077), Singapore, 23–27 January 2000; IEEE: Piscataway, NJ, USA, 2000; Volume 4, pp. 2417–2418.
9. Brignone, M.; Mestriner, D.; Procopio, R.; Piantini, A.; Rachidi, F. Evaluation of the mitigation effect of the shield wires on lightning induced overvoltages in mv distribution systems using statistical analysis. *IEEE Trans. Electromagn. Compat.* **2017**, *60*, 1400–1408. [CrossRef]
10. Nucci, C.A. A survey on Cigré and IEEE procedures for the estimation of the lightning performance of overhead transmission and distribution lines. In Proceedings of the 2010 Asia-Pacific International Symposium on Electromagnetic Compatibility, Beijing, China, 12–16 April 2010; IEEE: Piscataway, NJ, USA, 2010; pp. 1124–1133.
11. Napolitano, F.; Tossani, F.; Borghetti, A.; Nucci, C.A. Lightning performance assessment of power distribution lines by means of stratified sampling monte carlo method. *IEEE Trans. Power Deliv.* **2018**, *33*, 2571–2577. [CrossRef]
12. Nucci, C.A.; Guerrieri, S.; De Barros, M.C.; Rachidi, F. Influence of corona on the voltages induced by nearby lightning on overhead distribution lines. *IEEE Trans. Power Deliv.* **2000**, *15*, 1265–1273. [CrossRef]
13. Yu, Z.; Zhu, T.; Wang, Z.; Lu, G.; Zeng, R.; Lin, Y.; Luo, J.; Wang, Y.; He, J.; Zhuang, C. Calculation and experiment of induced lightning overvoltage on power distribution line. Progress on Lightning Research and Protection Technologies. *Electr. Power Syst. Res.* **2016**, *139*, 52–59. [CrossRef]

Appl. Sci. **2020**, *10*, 4902

14. Agrawal, A.K.; Price, H.J.; Gurbaxani, S.H. Transient Response of Multiconductor Transmission Lines Excited by a Nonuniform Electromagnetic Field. *IEEE Trans. Electromagn. Compat.* **1980**, *EMC-22*, 119–129. [CrossRef]

15. Noda, T. Development of A Transmission-Line Model Considering the Skin and Corona Effects for Power Systems Transient Analysis. Ph.D. Thesis, Doshisha University, Tokyo, Japan, 1996.

16. Thang, T.H.; Baba, Y.; Nagaoka, N.; Ametani, A.; Takami, J.; Okabe, S.; Rakov, V.A. A simplified model of corona discharge on overhead wire for FDTD computations. *IEEE Trans. Electromagn. Compat.* **2011**, *54*, 585–593. [CrossRef]

17. Thang, T.H.; Baba, Y.; Nagaoka, N.; Ametani, A.; Itamoto, N.; Rakov, V.A. FDTD simulations of corona effect on lightning-induced voltages. *IEEE Trans. Electromagn. Compat.* **2013**, *56*, 168–176. [CrossRef]

18. Brignone, M.; Mestriner, D.; Procopio, R.; Rossi, M.; Piantini, A.; Rachidi, F. EM Fields Generated by a Scale Model Helical Antenna and Its Use in Validating a Code for Lightning-Induced Voltage Calculation. *IEEE Trans. Electromagn. Compat.* **2019**, *61*, 778–787. [CrossRef]

19. Brignone, M.; Ginnante, E.; Mestriner, D.; Ruggi, L.; Procopio, R.; Piantini, A.; Rachidi, F. Evaluation of lightning-induced overvoltages on a distribution system: Validation of a dedicated code using experimental results on a reduced-scale model. In Proceedings of the 2017 IEEE International Conference on Environment and Electrical Engineering and 2017 IEEE Industrial and Commercial Power Systems Europe (EEEIC/I CPS Europe), Milan, Italy, 6–9 June 2017; pp. 1–6.

20. Cooray, G.V. *The Lightning Flash*; IET: Stevenage, UK, 2003.

21. Hartmann, G. Theoretical Evaluation of Peek's Law. *IEEE Trans. Ind. Appl.* **1984**, *IA-20*, 1647–1651. [CrossRef]

22. Bousiou, E.I.; Mikropoulos, P.N.; Zagkanas, V.N. Corona inception field of typical overhead line conductors under variable atmospheric conditions. *Electr. Power Syst. Res.* **2020**, *178*, 106032. [CrossRef]

23. Heidler, F.; Cvetic, J.M.; Stanic, B.V. Calculation of lightning current parameters. *IEEE Trans. Power Deliv.* **1999**, *14*, 399–404. [CrossRef]

24. Anderson, J. *Lightning Performance of Transmission Lines*; Electric Power Research Institute: Palo Alto, CA, USA, 1981.

25. Darveniza, M. A practical extension of Rusck's formula for maximum lightning-induced voltages that accounts for ground resistivity. *IEEE Trans. Power Deliv.* **2006**, *22*, 605–612. [CrossRef]

26. Kuffel, J.; Kuffel, P. *High Voltage Engineering Fundamentals*; Elsevier: Amsterdam, The Netherlands, 2000.

Article

An Efficient Method for Calculating the Lightning Electromagnetic Field Over Perfectly Conducting Ground

Xin Liu * and Tianping Ge

Department of Electrical Engineering, North China Electric Power University, Baoding 071003, China; gtp970321@163.com
* Correspondence: liuxinncepu@ncepu.edu.cn

Received: 29 May 2020; Accepted: 18 June 2020; Published: 22 June 2020

Abstract: In the implementation of the Cooray–Rubinstein formula, the calculation of a lightning electromagnetic field over perfectly conducting ground accounted for most of the computation time. Commonly, evaluating the ideal lightning electromagnetic field is based on the numerical integration method. In practice, only a sufficiently small discretization step is essential to get an accurate result, which leads to a relatively large number of calculations and results in a lengthy computation time. Besides, the programming is relatively complicated because the propagation of the lightning current along the channel must be considered. In order to increase the efficiency and simplify the programming, an improved method is proposed in this paper. In this method, the evaluation of the ideal lightning electromagnetic field is equated with a summation of analytical formulae and a simple integral operation, so it would be more efficient and easily programmed. The validation of the proposed method is demonstrated by some simulation examples.

Keywords: lightning; electromagnetic field; analytical formula

---

1. Introduction

As a prerequisite for evaluating lightning-induced voltages in distribution networks, it is of great significance to calculate the electromagnetic field generated by a lightning return stroke over a lossy ground accurately and efficiently. At present, there are two kinds of methods for the calculation: Finite-difference time-domain method (FDTD) and Cooray–Rubinstein (CR) formula. The FDTD method [1–3] has great applicability when considering of the effect of lossy ground, such as the lightning electromagnetic field over layered ground. However, the space and time are required to be discretized in FDTD, which results in a huge amount of computation time and relatively complicated programming, especially for the absorbing boundary. In order to reduce the computation time, analytical formulae are an alternative choice. The Sommerfeld integral [4] can be used to rigorously evaluate the lightning electromagnetic field generated by a lightning return stroke over a lossy ground; however, the highly oscillatory and slow convergent integrand make it difficult to evaluate the integral efficiently. The Cooray–Rubinstein formula [5–8] is a good choice to overcome this difficulty, and has become a widely used method. In the past years, most studies have mainly focused on efficient evaluation of the CR formula in the time domain [9–18]. Essentially, the main task of these methods is how to describe the integral kernel function of the CR formula in the time domain and accelerate the calculation of the convolution.

In practice, the implementation of the CR formula to evaluate the lightning electromagnetic field over a lossy ground requires a two-step procedure. The first step involves calculating the lightning electromagnetic field over a perfectly conducting ground, and the second step is to evaluate the CR formula. Actually, compared with the second step, the calculation of the ideal lightning electromagnetic field ($E_{r,ideal}$ and $H_{\varphi,ideal}$) accounted for most of the computation time. Therefore, increasing the efficiency of the first step is more conducive to improving the overall computational efficiency. The relevant studies about this tend to be neglected. The formulae for evaluating $E_{r,ideal}$ and $H_{\varphi,ideal}$ have been established based on the dipole method, and they are composed by integrals with respect to the lightning channel. The common method to evaluate these integrals are mainly based on the numerical integration, by means of a discretization of the lightning channel. However, only a sufficiently small discretization step is essential to get an accurate result, which leads to a relatively large number of calculations and results in a lengthy computation time. Besides, the programming is relatively complicated because the propagation of the lightning current along the channel must be considered.

In order to increase the efficiency and simplify the programming, an improved method is proposed in this paper. Firstly, regarding the return stroke current in the lightning channel as a series of current sources distributed along the channel, the original time-varying system can be transformed into a time-invariant system. Then, the calculation formulae are rewritten in the frequency domain by means of the Fourier transform, which simplifies the integral and differential operations about time in the original formula, so that the formulae are reduced from having two variables (z', t) in the time domain to having only one variable (z') in the frequency domain. Finally, a series of analytical formulae according to the integrals are derived, which effectively simplify the calculation procedure. Additionally, compared with the conventional method, the efficiency of the proposed method can be increased, which can be examined by the simulation example and discussion.

2. Method for Calculating the Lightning Electromagnetic Field over Perfectly Conducting Ground

2.1. Review of the Lightning Return Stroke Model

In order to calculate the lightning electromagnetic field, the lightning return stroke model must be established firstly. Basically, there are four kinds of lightning return stroke models: The gas dynamic or physical mode, the electromagnetic model, the distributed-circuit model, and the engineering model [19–22]. The engineering model is adopted here due to its wide utilization. There are several engineering models, such as the Bruce–Golde model (BG), transmission line model (TL), traveling current source model (TCS), modified transmission line exponential decay model (MTLE), and modified transmission line linear model (MTLL) [23]. In this paper, the MTLE model is adopted, which can be illustrated as Figure 1 and described by:

$$\begin{cases} i(z',t) = i(0, t - \frac{z'}{v}) \exp(-\frac{z'}{\alpha}), & t \geq \frac{z'}{v} \\ i(z',t) = 0, & t < \frac{z'}{v} \end{cases} \tag{1}$$

where: v—the return stroke velocity along the lightning channel, α—the decaying constant, $i(0, t - z'/v)$—the delay of the channel base current, and $\exp(-z'/\alpha)$—the attenuation of the channel base current.

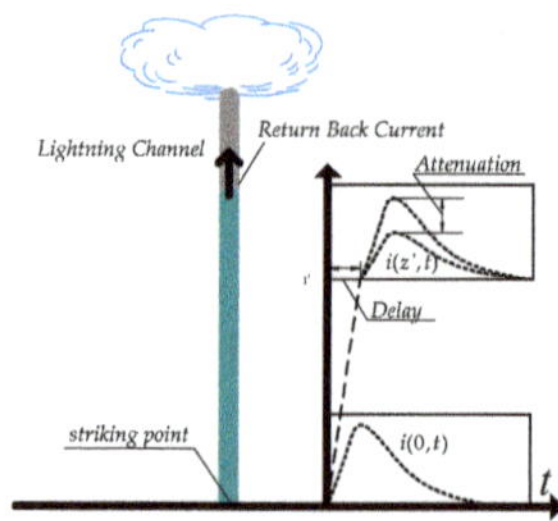

Figure 1. Lightning return stroke model.

As for the channel base current, several functions are available, such as Rubinstein and Uman function, Bruce and Golde function, and Heidler function [24]. In this paper, the channel base current is represented as a sum of two Heidler functions, as given in Equation (2), and the typical parameters are listed in Table 1:

$$i(0,t) = \left[\frac{i_{01}}{\eta_1} \frac{(t/\tau_{11})^{n_1}}{1+(t/\tau_{11})^{n_1}} \exp\left(-\frac{t}{\tau_{12}}\right) + \frac{i_{02}}{\eta_2} \frac{(t/\tau_{21})^{n_2}}{1+(t/\tau_{21})^{n_2}} \exp\left(-\frac{t}{\tau_{22}}\right) \right], \tag{2}$$

where:

$$\eta_1 = \exp\left(-\frac{\tau_{11}}{\tau_{12}}\left(n_1\frac{\tau_{12}}{\tau_{11}}\right)^{\frac{1}{n_1}}\right) \text{ and } \eta_2 = \exp\left(-\frac{\tau_{21}}{\tau_{22}}\left(n_1\frac{\tau_{22}}{\tau_{21}}\right)^{\frac{1}{n_2}}\right).$$

Table 1. The basic parameters for the channel base current function.

i_{01}	i_{02}	τ_{11}	τ_{12}	τ_{21}	τ_{22}	n_1	n_2
10.7×10^3 A	6.5×10^3 A	0.25×10^{-6} s	2.5×10^{-6} s	2×10^{-6} s	230×10^{-6} s	2	2

2.2. Fundamental Formulation of Lightning Electromagnetic Field over Perfectly Conducting Ground

The widely used model of the lightning electromagnetic field over a perfect conducting ground can be illustrated in Figure 2 [25]. The lightning channel can be regarded as a combination of an above-ground lightning channel and its mirror in the free space. Regarding the source as a vertical electric dipole of current $i(z',t)dz'$ located at z', the formulas of the electromagnetic field at an observation point generated by the lightning channel can be obtained and shown as:

$$\begin{cases} E_{ideal} = \frac{1}{4\pi\varepsilon_0} \left\{ \int_0^h \left[\begin{array}{l} \frac{3r(z-z')}{R_{z'}^5}\int_0^t i(z',\tau - R_{z'}/c)d\tau \\ +\frac{3r(z-z')}{cR_{z'}^4}i(z',t-R_{z'}/c) \\ +\frac{r(z-z')}{c^2R_{z'}^3}\frac{\partial i(z',t-R_{z'}/c)}{\partial t} \end{array} \right] dz' + \int_0^h \left[\begin{array}{l} \frac{3r(z+z')}{R_{-z'}^5}\int_0^t i(z',\tau - R_{-z'}/c)d\tau \\ +\frac{3r(z+z')}{cR_{-z'}^4}i(z',t-R_{-z'}/c) \\ +\frac{r(z+z')}{c^2R_{-z'}^3}\frac{\partial i(z',t-R_{-z'}/c)}{\partial t} \end{array} \right] dz' \right\} \\ H_{ideal} = \frac{1}{4\pi}\left\{ \int_0^h\left[\frac{r}{R_{z'}^3}i(z',t-R_{z'}/c) + \frac{r}{cR_{z'}^2}\frac{\partial i(z',t-R_{z'}/c)}{\partial t} \right]dz' \\ +\int_0^h\left[\frac{r}{R_{-z'}^3}i(z',t-R_{-z'}/c) + \frac{r}{cR_{-z'}^2}\frac{\partial i(z',t-R_{-z'}/c)}{\partial t} \right]dz' \right\} \end{cases}, \tag{3}$$

where: $E_{ideal}(\varphi, r, z, t)$—ideal horizontal electric field at observation point $P(r, z)$, $H_{ideal}(\varphi, r, z, t)$—ideal tangential magnetic field at observation point $P(r, z)$, h—the height of the lightning channel, ε_0—vacuum permittivity, r—horizontal distance from the lightning channel to point $P(r, z)$, $R_{z'}$ and $R_{-z'}$—distance from the dipole to point $P(r, z)$, z and z'—the z-coordinate of point $P(r, z)$ and the dipole, respectively, and c—speed of light in vacuum.

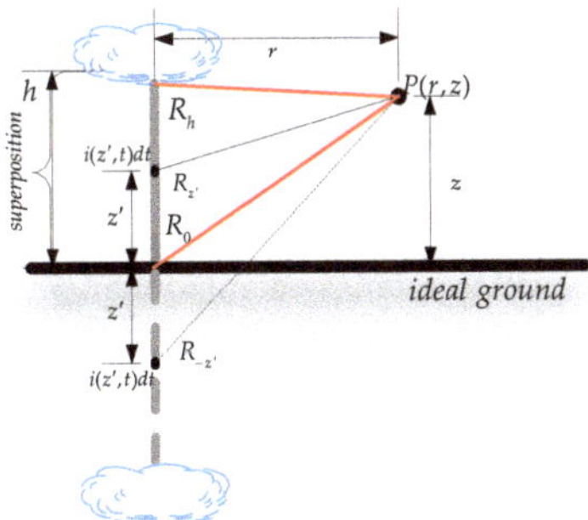

Figure 2. Model for calculating the lightning electromagnetic field.

The calculation of Equation (3) is commonly based on a numerical integration due to the lack of the analytical formulae of Equation (3), which is called the conventional method in this paper. As can be seen, the integrands in Equation (3) are functions of two variables (t and z'), and the propagation of the current along the channel must be considered when implementing the numerical integration. Therefore, it is somewhat complicated and not easily programmed. Experience indicates that a sufficiently small computational step dz' is essential to get an accurate result, but this leads to a large number of calculations, resulting in a lengthy calculation time.

Therefore, we doubt whether or not some of the integral can be implemented in an analytical way; if so, the efficiency can be improved, especially for the evaluation of the lighting-induced voltages of a distributed network, in which lightning electromagnetic fields with a great amount of observation points are required to be calculated. In the following sections, an efficient method is achieved, which can calculate most of the integrals in Equation (3) analytically.

2.3. Proposed Method for Calculating the Lightning Electromagnetic Field over Perfectly Conducting Ground

In order to simplify Equation (3), firstly, we transform the system from time varying to time invariant. The process of the channel base current propagating along the lightning channel is time varying, but if regarded it as a current model with $i(z', t) = i_{base}(t-z'/v)$ distribution at all points along the channel direction, the system can be transformed into a time-invariant system. Then, we can transform the time-invariant system into the frequency domain based on the Fourier transform. Finally, the analytical formulae for most of the integral in Equation (3) can be obtained by means of some derivations.

Substituting Equations (1) and (2) into (3) and representing the equations in the frequency domain allows the formulas to be rearranged as:

$$E_{ideal} = \frac{1}{4\pi\varepsilon_0}\left\{ \int_0^{h} \begin{bmatrix} \frac{3r(z-z')}{R_{z'}^5}\frac{I(0,j\omega)}{j\omega}e^{-z'(\frac{j\omega}{v}+\frac{1}{a})}e^{-jkR_{z'}} \\ +\frac{3r(z-z')}{cR_{z'}^4}I(0,j\omega)e^{-z'(\frac{j\omega}{v}+\frac{1}{a})}e^{-jkR_{z'}} \\ +j\omega\frac{r(z-z')}{c^2R_{z'}^3}I(0,j\omega)e^{-z'(\frac{j\omega}{v}+\frac{1}{a})}e^{-jkR_{z'}} \end{bmatrix}dz' \underbrace{}_{E_{above}} + \int_0^{h} \begin{bmatrix} \frac{3r(z+z')}{R_{-z'}^5}\frac{I(0,j\omega)}{j\omega}e^{-z'(\frac{j\omega}{v}+\frac{1}{a})}e^{-jkR_{-z'}} \\ +\frac{3r(z+z')}{cR_{-z'}^4}I(0,j\omega)e^{-z'(\frac{j\omega}{v}+\frac{1}{a})}e^{-jkR_{-z'}} \\ +j\omega\frac{r(z+z')}{c^2R_{-z'}^3}I(0,j\omega)e^{-z'(\frac{j\omega}{v}+\frac{1}{a})}e^{-jkR_{-z'}} \end{bmatrix}dz' \underbrace{}_{E_{under}} \right\} \quad (4)$$

As can be seen in Equation (4), E_{above} and E_{under} have the same form, so only the derivation for E_{above} is provided. The above-ground lightning channel-generated horizontal electric field can be divided into three parts, i.e., E_A, E_B, and E_C, as shown in Equation (5):

$$E_{above} = \frac{I(0,j\omega)r}{4\pi\varepsilon_0}\left\{ \underbrace{\frac{3}{j\omega}\int_0^{h}\frac{(z-z')}{R_{z'}^5}\cdot e^{-z'(\frac{j\omega}{v}+\frac{1}{a})}e^{-jkR_{z'}}dz'}_{E_A} + \underbrace{\frac{3}{c}\int_0^{h}\frac{(z-z')}{R_{z'}^4}\cdot e^{-z'(\frac{j\omega}{v}+\frac{1}{a})}e^{-jkR_{z'}}dz'}_{E_B} + \underbrace{j\omega\frac{1}{c^2}\int_0^{h}\frac{(z-z')}{R_{z'}^3}\cdot e^{-z'(\frac{j\omega}{v}+\frac{1}{a})}e^{-jkR_{z'}}dz'}_{E_C} \right\},\left\{k=\frac{\omega}{c}\right\} \quad (5)$$

By means of $\int\frac{(z-z')}{R_{z'}^5}dz' = \frac{1}{3R_{z'}^3}$ and $\int u'vdz = uv - \int uv'dz$, E_A can be rewritten as:

$$E_A = \left\{ \underbrace{\frac{1}{j\omega}\cdot\frac{1}{R_{z'}^3}e^{-z'(\frac{j\omega}{v}+\frac{1}{a})}e^{-jkR_{z'}}\Big|_0^{h}}_{E_{A1}} + \underbrace{\frac{1}{j\omega}\cdot\left(\frac{j\omega}{v}+\frac{1}{a}\right)\int_0^{h}\frac{1}{R_{z'}^3}e^{-z'(\frac{j\omega}{v}+\frac{1}{a})}e^{-jkR_{z'}}dz'}_{E_{A2}} + \underbrace{\left(-\frac{1}{c}\int_0^{h}\frac{(z-z')}{R_{z'}^4}e^{-z'(\frac{j\omega}{v}+\frac{1}{a})}e^{-jkR_{z'}}dz'\right)}_{E_{A3}} \right\} \quad (6)$$

in which E_{A2} can be described as Equation (7) based on $\int \frac{(z-z')}{R_{z'}^3} dz' = \frac{1}{R_{z'}}$ and $\int u'v\, dz = uv - \int uv'\, dz$:

$$
E_{A2} = \left\{
\begin{array}{l}
\underbrace{\frac{1}{j\omega} \cdot \left(\frac{j\omega}{v} + \frac{1}{\alpha}\right) \frac{1}{R_{z'}} \frac{1}{z-z'} e^{-z'\left(\frac{j\omega}{v}+\frac{1}{\alpha}\right)} e^{-jkR_{z'}} \bigg|_0^h}_{E_{A21}}
+ \underbrace{\left(-\frac{1}{j\omega} \cdot \left(\frac{j\omega}{v} + \frac{1}{\alpha}\right) \int_0^h \frac{1}{R_{z'}} \frac{1}{(z-z')^2} e^{-z'\left(\frac{j\omega}{v}+\frac{1}{\alpha}\right)} e^{-jkR_{z'}} dz'\right)}_{E_{A22}} \\[2em]
\underbrace{+\frac{1}{j\omega} \cdot \left(\frac{j\omega}{v} + \frac{1}{\alpha}\right)\left(\frac{j\omega}{v} + \frac{1}{\alpha}\right) \int_0^h \frac{1}{R_{z'}} \frac{1}{z-z'} e^{-z'\left(\frac{j\omega}{v}+\frac{1}{\alpha}\right)} e^{-jkR_{z'}} dz'}_{E_{A23}}
+ \underbrace{\left(-\frac{1}{c} \cdot \left(\frac{j\omega}{v} + \frac{1}{\alpha}\right) \int_0^h \frac{1}{R_{z'}^2} e^{-z'\left(\frac{j\omega}{v}+\frac{1}{\alpha}\right)} e^{-jkR_{z'}} dz'\right)}_{E_{A24}}
\end{array}
\right\}
\tag{7}
$$

As for E_{A22}, according to $\int \frac{1}{R_{z'}(z-z')^2} dz' = \frac{R_{z'}}{r^2(z-z')}$ and $\int u'v\, dz = uv - \int uv'\, dz$, it will be:

$$
E_{A22} = \left\{
\begin{array}{l}
\underbrace{-\frac{1}{j\omega} \cdot \left(\frac{j\omega}{v} + \frac{1}{\alpha}\right) \frac{R_{z'}}{r^2(z-z')} e^{-z'\left(\frac{j\omega}{v}+\frac{1}{\alpha}\right)} e^{-jkR_{z'}} \bigg|_0^h}_{E_{A221}}
+ \underbrace{\frac{1}{j\omega} \cdot \left(\frac{j\omega}{v} + \frac{1}{\alpha}\right) jk\frac{1}{r^2} \int_0^h e^{-z'\left(\frac{j\omega}{v}+\frac{1}{\alpha}\right)} e^{-jkR_{z'}} dz'}_{E_{A222}} \\[2em]
\underbrace{+\left(-\frac{1}{j\omega} \cdot \left(\frac{j\omega}{v} + \frac{1}{\alpha}\right)\left(\frac{j\omega}{v} + \frac{1}{\alpha}\right) \int_0^h \frac{R_{z'}}{r^2(z-z')} e^{-z'\left(\frac{j\omega}{v}+\frac{1}{\alpha}\right)} e^{-jkR_{z'}} dz'\right)}_{E_{A223}}
\end{array}
\right\}
\tag{8}
$$

Combining E_{A3} with E_B and using $\int \frac{(z-z')}{R_{z'}^4} dz' = \frac{1}{2R_{z'}^2}$, we can get:

$$
E_{A3} + E_B = \left\{
\begin{array}{l}
\underbrace{\frac{2}{c} \frac{1}{2R_{z'}^2} e^{-z'\left(\frac{j\omega}{v}+\frac{1}{\alpha}\right)} e^{-jkR_{z'}} \bigg|_0^h}_{E_{B1}}
+ \underbrace{\frac{1}{c}\left(\frac{j\omega}{v} + \frac{1}{\alpha}\right) \int_0^h \frac{1}{R_{z'}^2} e^{-z'\left(\frac{j\omega}{v}+\frac{1}{\alpha}\right)} e^{-jkR_{z'}} dz'}_{E_{B2}} \\[2em]
\underbrace{+\left(-j\omega\frac{1}{c^2} \int_0^h \frac{(z-z')}{R_{z'}^3} e^{-z'\left(\frac{j\omega}{v}+\frac{1}{\alpha}\right)} e^{-jkR_{z'}} dz'\right)}_{E_{B3}}
\end{array}
\right\}.
\tag{9}
$$

E_{A222}, E_{A223}, and E_{A23} can also be combined, and the summation of them can be expressed as Equation (10), taking advantage of $\int_0^h \frac{z-z'}{R_{z'}} e^{-jkR_{z'}} dz' = \frac{e^{-jkR_{z'}}}{jk}$:

$$
E_{A222} + E_{A223} + E_{A23} = \left\{
\begin{array}{l}
\underbrace{-\frac{1}{j\omega r^2} \cdot \left(\frac{j\omega}{v} + \frac{1}{\alpha}\right)\left(\frac{j\omega}{v} + \frac{1}{\alpha}\right) \frac{e^{-jkR_{z'}}}{jk} e^{-z'\left(\frac{j\omega}{v}+\frac{1}{\alpha}\right)} \bigg|_0^h}_{E_{A231}} \\[2em]
\underbrace{+\frac{1}{j\omega r^2}\left(\frac{j\omega}{v} + \frac{1}{\alpha}\right)\left[jk - \frac{1}{jk}\left(\frac{j\omega}{v} + \frac{1}{\alpha}\right)^2\right] \int_0^h e^{-z'\left(\frac{j\omega}{v}+\frac{1}{\alpha}\right)} e^{-jkR_{z'}} dz'}_{E_{A241}}
\end{array}
\right\}.
\tag{10}
$$

Considering $E_{B2} + E_{A24} = 0$ and $E_{B3} + E_C = 0$, Equation (5) can be rewritten as:

$$
E_{above} = \frac{I(0,j\omega)r}{4\pi\varepsilon_0}\left\{
\begin{aligned}
&\underbrace{\frac{1}{j\omega}\cdot\frac{1}{R_{z'}^3}e^{-z'(\frac{j\omega}{v}+\frac{1}{\alpha})}e^{-jkR_{z'}}\bigg|_0^h}_{E_{A1}} + \underbrace{\frac{2}{c}\frac{1}{2R_{z'}^2}e^{-z'(\frac{j\omega}{v}+\frac{1}{\alpha})}e^{-jkR_{z'}}\bigg|_0^h}_{E_{B1}} \\
&+\underbrace{\frac{1}{j\omega}\cdot\left(\frac{j\omega}{v}+\frac{1}{\alpha}\right)\frac{1}{R_{z'}}\frac{1}{z-z'}e^{-z'(\frac{j\omega}{v}+\frac{1}{\alpha})}e^{-jkR_{z'}}\bigg|_0^h}_{E_{A21}} \\
&+\underbrace{\left(-\frac{1}{j\omega}\cdot\left(\frac{j\omega}{v}+\frac{1}{\alpha}\right)\frac{R_{z'}}{r^2(z-z')}e^{-z'(\frac{j\omega}{v}+\frac{1}{\alpha})}e^{-jkR_{z'}}\bigg|_0^h\right)}_{E_{A221}} \\
&+\underbrace{\left(-\frac{1}{j\omega r^2}\cdot\left(\frac{j\omega}{v}+\frac{1}{\alpha}\right)\left(\frac{j\omega}{v}+\frac{1}{\alpha}\right)\frac{e^{-jkR_{z'}}}{jk}e^{-z'(\frac{j\omega}{v}+\frac{1}{\alpha})}\bigg|_0^h\right)}_{E_{A231}} \\
&+\underbrace{\left[j\omega\frac{1}{v}\left(\frac{1}{c}-\frac{c}{v^2}\right)+\frac{1}{\alpha}\left(\frac{1}{c}-\frac{3c}{v^2}\right)-\frac{1}{j\omega}\frac{3c}{\alpha^2 v}-\frac{1}{(j\omega)^2}\frac{c}{\alpha^3}\right]\cdot\frac{1}{r^2}\int_0^h e^{-z'(\frac{j\omega}{v}+\frac{1}{\alpha})}e^{-jkR_{z'}}\,dz'}_{E_{A241}}
\end{aligned}
\right\}
\tag{11}
$$

and it can be transformed into the time domain, i.e.:

$$
E_{above} = E_{arithmetic} + E_{integral},
\tag{12}
$$

where:

$$
E_{arithmetic} = \frac{1}{4\pi\varepsilon_0}\left\{
\begin{aligned}
&\left(\frac{r}{c}\frac{1}{R_h^2}+\frac{r}{vR_h}\frac{1}{z-h}-\frac{R_h}{vr(z-h)}-\frac{c}{rv^2}\right)i\left(0,t-\frac{h}{v}-\frac{R_h}{c}\right)e^{-\frac{h}{\alpha}} \\
&-\left(\frac{r}{c}\frac{1}{R_0^2}+\frac{r}{vR_0}\frac{1}{z}-\frac{R_0}{vrz}-\frac{c}{rv^2}\right)i\left(0,t-\frac{R_0}{c}\right) \\
&+\left(\frac{r}{R_h^3}+\frac{1}{\alpha}\frac{r}{R_h}\frac{1}{z-h}-\frac{1}{\alpha}\frac{R_h}{r(z-z')}-\frac{2c}{\alpha vr}\right)i^{-1}\left(0,t-\frac{h}{v}-\frac{R_h}{c}\right)e^{-\frac{h}{\alpha}} \\
&-\left(\frac{r}{R_0^3}+\frac{1}{\alpha}\frac{r}{R_0}\frac{1}{z}-\frac{1}{\alpha}\frac{R_0}{rz}-\frac{2c}{\alpha vr}\right)i^{-1}\left(0,t-\frac{R_0}{c}\right) \\
&-\frac{c}{r\alpha^2}i^{-2}\left(0,t-\frac{h}{v}-\frac{R_h}{c}\right)e^{-\frac{h}{\alpha}}+\frac{c}{r\alpha^2}i^{-2}\left(0,t-\frac{R_0}{c}\right)
\end{aligned}
\right\}
\tag{13}
$$

$$
E_{integral} = \left\{
\begin{aligned}
&\frac{1}{v}\left(\frac{1}{c}-\frac{c}{v^2}\right)\cdot\frac{1}{r}\cdot\frac{d}{dt}i_{integral}(t)+\frac{1}{\alpha}\left(\frac{1}{c}-\frac{3c}{v^2}\right)\cdot\frac{1}{r}\cdot i_{integral}(t) \\
&-\frac{3c}{\alpha^2 v}\cdot\frac{1}{r}\cdot i^{-1}_{integral}(t)-\frac{c}{\alpha^3}\cdot\frac{1}{r}\cdot i^{-2}_{integral}(t)
\end{aligned}
\right\},
\tag{14}
$$

where: $i(0,t)$—the channel base current, $i\left(0,t-\frac{h}{v}-\frac{R_h}{c}\right)$ and $i\left(0,t-\frac{R_0}{c}\right)$—the channel base current with delay, $i_{integral}(t) = \frac{1}{4\pi\varepsilon_0}\int_0^h i\left(0,t-\frac{z'}{v}-\frac{R_{z'}}{c}\right)e^{-z'(\frac{1}{\alpha})}\,dz'$—the integral component associated with the channel base current, R_h—the distance between the highest point of the lightning channel and point $P(r,z)$, and R_0—the distance between the lightning strike point and point $P(r,z)$.

By means of the similar derivation, the formula for E_{under} can also be achieved, only by replacing $z-z'$ and $R_{z'}$ by $z+z'$ and $R_{-z'}$, respectively. Finally, the horizontal electric field E_{ideal} can be obtained by adding them together.

As can be seen from Equations (12)–(14), almost all of the integral can be evaluated in an analytical way, except for $E_{integral}$. Moreover, the numerical calculation of $E_{integral}$ is very simple, because it only contains the integral of the current along the channel and its derivatives. Generally, comparing Equations (12)–(14) with Equation (3), it can be seen that the proposed method equates the integral operation with a simple arithmetic operation and a simple integral. The arithmetic operation involves fewer steps, requires less data, and produces more accurate results. The integral is much simpler than that of the conventional method. Thus, doing this makes the proposed method more efficient and easily programmed.

3. Results

In order to demonstrate the superiority of the proposed method, some simulations were performed in MATLAB on a PC with i7 CPU. To focus on the difference between the two methods, the discretization step along the lightning channel, dz', was adopted as the variable, while the other parameters are listed in Table 2. The main purpose of this paper was to deal with the calculation problem of the horizontal electric field and it is noted that the horizontal electric field has great attenuation when it is more than 1 km away from the lightning return channel, so the lightning striking distance r is restricted to 0~1 km. According to the geometric formula, the height difference caused by the curvature of the earth is below 8 cm, which is very small compared with the distance of 1 km [26], so the influence of the curvature of the earth is ignored in this paper.

Table 2. Basic parameters used in the simulation examples.

t	v	r	z	c	α	dt	ε_0
30 μs	1.3×10^8 m/s	50/250/500 m	10 m	3×10^8 m/s	1700 m	10 ns	8.854×10^{-12}

Since both of the methods include numerical integration with respect to dz', the value of dz' would inevitably influence the accuracy. Theoretically, the smaller the value of dz', the more accurate the result. However, there must be a trade-off between the adoption of dz' and the efficiency.

Firstly, the effect of dz' on the calculation accuracy of the two methods was studied, which is shown in Figure 3. By comparing Figure 3a with b, it can be seen that dz' has a great influence on the horizontal electric field's calculation of the conventional method, while it has basically no influence on that of the proposed method. When $dz' \geq 0.05$ m, the horizontal electric field obtained by the conventional method will deviate from the real value. However, no matter what value of dz' is taken by the proposed method, the final calculation results are consistent. Therefore, the proposed method is more accurate when the dz' is the same.

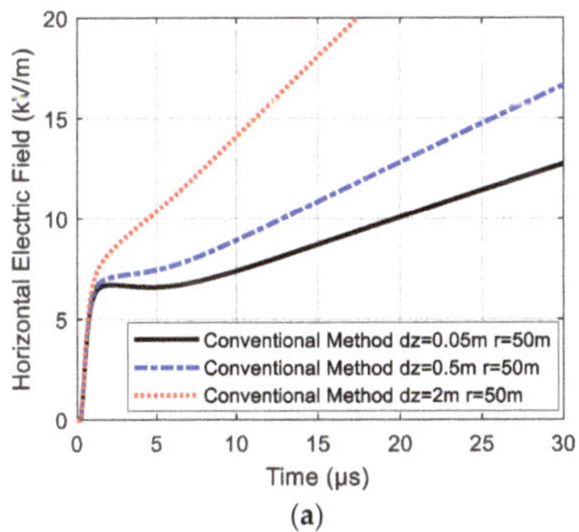

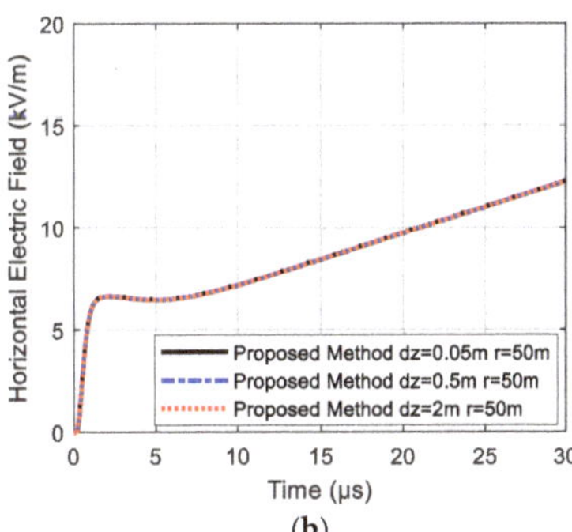

Figure 3. The time-domain evolutions of the horizontal electric field at the observation point under different dz'; (**a**) calculated by the conventional method; (**b**) calculated by the proposed method.

The calculation formulae of the proposed method are composed of two parts: The arithmetic part as shown in Equation (13) and the integral part as shown in Equation (14). In order to better explain the reason why the proposed method is not sensitive to dz', these two parts were analyzed respectively. For the arithmetic part, it is an error-free analytical result completely independent of dz' in the proposed method (analytical calculation) while an inexact numerical result sensitive to dz' in the conventional method (numerical calculation), as shown in Figure 4a,b. This indicates that the proposed method is more accurate in this part than the conventional method. For the integral part, its accuracy would be dependent on dz'. However, as can be seen in Figure 4c, the results are coincident together, although dz' takes different values. That is to say, an adoption of $dz' = 2$ m is enough to achieve the accurate result. This is why the proposed method can guarantee the accuracy with a relatively larger dz'.

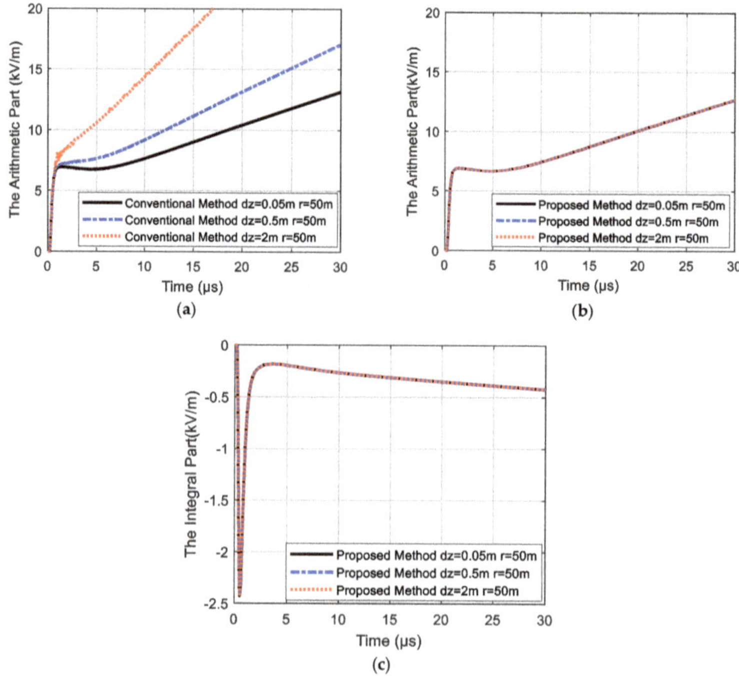

Figure 4. The two parts of the horizontal electric field at the observation point: (**a**) the arithmetic parts calculated by the conventional method; (**b**) the arithmetic parts calculated by the proposed method; (**c**) the integral part calculated by the proposed method.

The most fundamental reason why the proposed method can improve the efficiency can be well explained by Figure 5. It can be seen that the horizontal electric field calculated by the proposed method under a condition of $dz' = 2$ m is close to that of the conventional method with $dz' = 0.05$ m, and the computation time is 0.143 and 5.776 s, respectively, as shown in Table 3. That is to say, the acceleration of the proposed method is more efficient than that of the conventional method with the same accuracy.

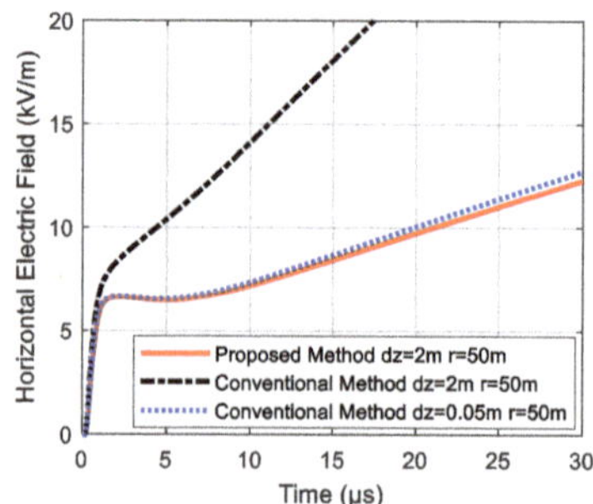

Figure 5. The horizontal electric field curves at the observation point calculated by the proposed method and the conventional method under different dz'.

Table 3. Comparison of the computation time between two methods.

dz'	Conventional Method	Proposed Method
0.05 m	5.776 s	-
2 m	0.121 s	0.143 s

Considering the calculation of lightning-induced voltages of distributed overhead lines, it is often necessary to calculate a large number of observation points. Taking a one-kilometer three-phase line network as an example, it is generally necessary to calculate at least 300 observation points. According to Table 3, the computation time of the proposed method will be about 1 min. It can be imagined that the more complex the power networks studied, the more efficient the proposed method. Additionally, compared with the conventional method, the proposed method is easier to be programmed because the only formula required to be calculated numerically is very simple.

We also considered the cases of r = 250 and 500 m. In these examples, the static E-field component of the horizontal electric field will gradually decrease, but the advantages of the proposed method still remain. The comparison of the two methods is shown in Figure 6 and Table 4. It can be seen from Figure 6b that the horizontal electric field calculated by the conventional method with dz' = 2m is close to the one with dz' = 0.05 m when r reaches more than 500 m. The efficiency of the two methods is almost the same, as the computation time listed in Table 4. However, as for the relative closer observation points, such as r = 250 m, small dz' is still required, which can be examined in Figure 6a. In a word, as for the conventional method, small dz' should be adopted for the observation point near the lighting channel, while large dz' can be used for the observation point far away. However, in the proposed method, small dz' can be adopted no matter the distance from the observation point and the lightning striking point, which indicates that the proposed method is more general. In other words, the proposed method can adopt uniform dz' without considering the different lightning striking distances. Therefore, its implementation is more convenient. Additionally, as the statement above, the proposed method is easier to be programmed because the formulation of the item for numerical integration is simpler, which is superior to the proposed method.

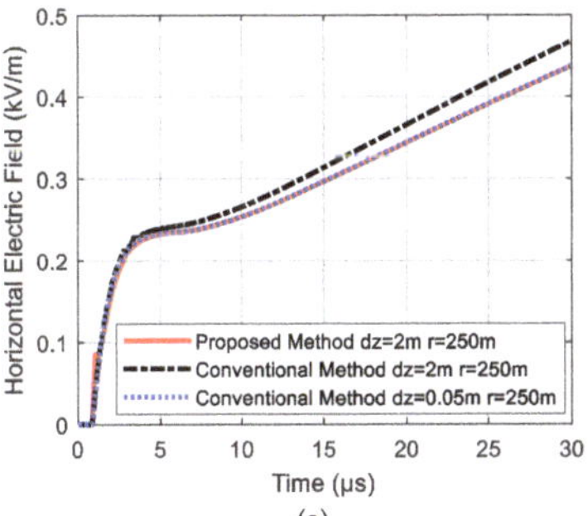

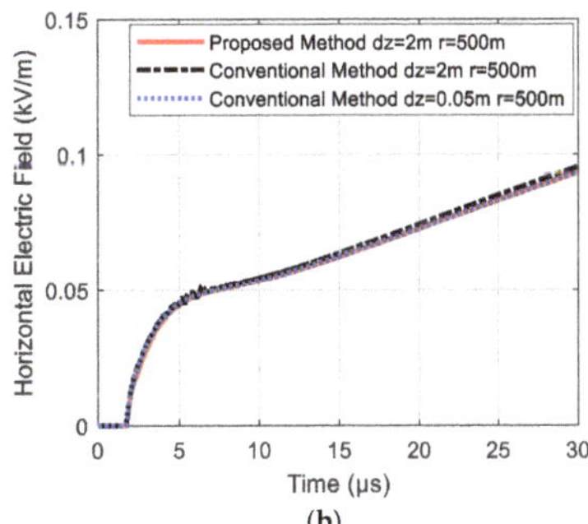

Figure 6. The horizontal electric field curves at the different observation points calculated by the proposed method and the conventional method under different dz'. (**a**) the horizontal electric field at r = 250 m calculated by two methods; (**b**) the horizontal electric field at r = 500 m calculated by two methods.

Table 4. Comparison of the computation time between the two methods.

dz'	Conventional Method		Proposed Method	
-	r = 250 m	r = 500 m	r = 250 m	r = 500 m
0.05 m	4.688 s	4.673 s	-	-
2 m	0.203 s	0.163 s	0.150 s	0.142 s

It should also be pointed out that the horizontal component of the lightning electric field attenuates greatly with the distance, as shown in Figures 4–6. Considering the lightning striking distances of more than 500 m, the overvoltage induced by the lightning electromagnetic field on the power line is generally small, which poses little threat to the insulation. Therefore, the example of the horizontal component of the lightning electric field at a longer distance is not discussed in this paper.

As for the calculation of the magnetic field near the lightning channel, its formula can be derived in the same manner as the horizontal electric field. It is not described in this paper for the sake of brevity.

4. Discussion and Conclusions

For the calculation of lightning electromagnetic fields over a perfectly conducting ground, the common method is to evaluate the integral by means of numerical integration. In the proposed method, the formula for calculating the lightning electromagnetic field is divided into two parts: One that can be solved analytically, and the other that can be solved numerically only by integral operations.

For the conventional method, the results are sensitive to dz' at a close distance (tens of meters), which must be small enough to get an accurate numerical solution. However, when dz' is too small, the amount of data required in the program is huge and the calculation steps are tedious, resulting in a lengthy time and slow calculation speed. As for the proposed method, dz' can be adopted as a relatively large value, so the computational efficiency can be improved inevitably. Two calculation methods were simulated in MATLAB, of which the simulation results showed that under the same accuracy, the calculation time of the proposed method is about 1/40 of that of the original method. As for large lightning striking distances (hundreds of meters), the conventional method's results are not sensitive to dz' anymore, and relative larger dz' can be adopted to get an accurate numerical solution. In other words, for the conventional method, small dz' should be adopted for the observation point near the lighting channel while large dz' can be used for the observation point far away. However, the proposed method has better generality because it can calculate the horizontal electric fields at different lightning striking distances with a uniform dz', which makes its implementation more convenient and easily programmed. Additionally, as can be seen, the only item required to be calculated numerically is very simple and other items can be implemented in an analytical way, so the proposed method is very easy to be programmed, compared with the original method.

Author Contributions: Conceptualization, X.L.; methodology, X.L. and T.G.; software, X.L. and T.G.; validation, X.L. and T.G.; formal analysis, T.G. and X.L.; writing—original draft preparation, T.G.; writing—review and editing, X.L.; visualization, T.G.; supervision, X.L.; project administration, X.L. All authors have read and agreed to the published version of the manuscript.

Funding: This research received no external funding.

Conflicts of Interest: The authors declare no conflict of interest.

References

1. Baba, Y.; Rakov, V.A. Applications of the FDTD Method to Lightning Electromagnetic Pulse and Surge Simulations. *IEEE Trans. Electromagn. Compat.* **2014**, *56*, 1506–1521. [CrossRef]
2. Yang, C.; Zhou, B. Calculation methods of electromagnetic fields very close to lightning. *IEEE Trans. Electromagn. Compat.* **2004**, *46*, 133–141. [CrossRef]
3. Sartori, C.A.F.; Cardoso, J.R. An analytical-FDTD method for near LEMP calculation. *IEEE Trans. Magn.* **2000**, *36*, 1631–1634. [CrossRef]
4. Sommerfeld, A. *Partial Differential Equation in Physics*; Academic Press: New York, NY, USA, 1949.
5. Cooray, V. Horizontal fields generated by return strokes. *Radio Sci.* **1992**, *27*, 529–537. [CrossRef]
6. Rubinstein, M. An approximate formula for the calculation of the horizontal electric field from lightning at close, intermediate, and long range. *IEEE Trans. Electromagn. Compat.* **1996**, *38*, 531–535. [CrossRef]
7. Cooray, V.; Scuka, V. Lightning-induced overvoltage in power lines: Validity of various approximations made in overvoltage calculations. *IEEE Trans. Electromagn. Compat.* **1998**, *40*, 355–363. [CrossRef]

8. Cooray, V. Some considerations on the "Cooray-Rubinstein" formulation used in deriving the horizontal electric field of lightning return strokes over finitely conducting ground. *IEEE Trans. Electromagn. Compat.* **2002**, *44*, 560–566. [CrossRef]

9. Caligaris, C.; Delfino, F.; Procopio, R. Cooray–Rubinstein Formula for the Evaluation of Lightning Radial Electric Fields: Derivation and Implementation in the Time Domain. *IEEE Trans. Electromagn. Compat.* **2008**, *50*, 194–197. [CrossRef]

10. Andreotti, A.; Rachidi, F.; Verolino, L. On the Kernel of the Cooray–Rubinstein Formula in the Time Domain. *IEEE Trans. Electromagn. Compat.* **2016**, *58*, 927–930. [CrossRef]

11. Zou, J.; Li, M.; Li, C.; Lee, J.B.; Chang, S.H. Fast Evaluation of the Horizontal Electric Field of Lightning with Cooray–Rubinstein Formula in Time Domain Using a Piecewise Quadratic Convolution Technique. *IEEE Trans. Electromagn. Compat.* **2012**, *54*, 1034–1041. [CrossRef]

12. Liu, X.; Yang, J.; Wang, L.; Liang, G. A New Method for Fast Evaluation of the Cooray–Rubinstein Formula in Time Domain. *IEEE Trans. Electromagn. Compat.* **2017**, *59*, 1188–1195. [CrossRef]

13. Liu, X.; Zhang, M.; Yang, J.; Wang, L.; Wang, C. Fast evaluation of the Cooray–Rubinstein formula in time domain via piecewise recursive convolution technique. *IET Gener. Transm. Distrib.* **2017**, *12*, 1404–1410. [CrossRef]

14. Andreotti, A.; Rachidi, F.; Verolino, L. A New Formulation of the Cooray–Rubinstein Expression in Time Domain. *IEEE Trans. Electromagn. Compat.* **2015**, *57*, 391–396. [CrossRef]

15. Andreotti, A.; Rachidi, F.; Verolino, L. Some Developments of the Cooray–Rubinstein Formula in the Time Domain. *IEEE Trans. Electromagn. Compat.* **2015**, *57*, 1079–1085. [CrossRef]

16. Andreotti, A.; Rachidi, F.; Verolino, L. A New Solution for the Evaluation of the Horizontal Electric Fields from Lightning in Presence of a Finitely Conducting Ground. *IEEE Trans. Electromagn. Compat.* **2018**, *60*, 674–678. [CrossRef]

17. Barbosa, C.F.; Paulino, J.O.S. On Time-Domain Expressions for Calculating the Horizontal Electric Field from Lightning. *IEEE Trans. Electromagn. Compat.* **2019**, *61*, 434–439. [CrossRef]

18. Liu, X.; Zhang, M.; Yang, J. Fast Evaluation of the Lightning Horizontal Electric Field over a Lossy Ground Based on the Time-Domain Cooray–Rubinstein Formula. *IEEE Trans. Electromagn. Compat.* **2019**, *61*, 449–457. [CrossRef]

19. Plooster, M.N. Numerical model of the return stroke of the lightning discharge. *Phys. Fluids* **1971**, *14*, 2124–2133. [CrossRef]

20. Moini, R.; Kordi, B.; Rafi, G.Z.; Rakov, V.A. A new lightning return stroke model based on antenna theory. *J. Geophys. Res.* **2000**, *105*, 29693–29702. [CrossRef]

21. Gorin, B.N.; Markin, V.I. Lightning return stroke as a transient process in a distributed system. *Trudy ENIN* **1975**, *43*, 114–130.

22. Rakov, V.; Dulzon, A. A Modified Transmission Line Model for Lightning Return Stroke Field Calculations. In Proceedings of the 9th International Symposium Electromagnetic Compatibility, Zurich, Switzerland, 12–14 March 1991; pp. 229–235.

23. Nucci, C.A.; Diendorfer, G.; Uman, M.A.; Rachidi, F.; Ianoz, M.; Mazzetti, C. Lightning return stroke current models with specified channel-base current: A review and comparison. *J. Geophys. Res.* **1990**, *95*, 20395–20408. [CrossRef]

24. Bruce, C.E.R.; Golde, R.H. *The Lightning Discharge*; Courier Corporation: Chelmsford, MA, USA, 2001.

25. Rubinstein, M.; Uman, M.A. Transient electric and magnetic fields associated with establishing a finite electrostatic dipole, revisited. *IEEE Trans. Electromagn. Compat.* **1991**, *33*, 312–320. [CrossRef]

26. Watson, G.N. The Transmission of Electric Waves Round the Earth. *Proc. R. Soc. Lond.* **1919**, *95*, 546–563.

Article

On the Efficiency of OpenACC-aided GPU-Based FDTD Approach: Application to Lightning Electromagnetic Fields

Sajad Mohammadi [1], Hamidreza Karami [1,*], Mohammad Azadifar [2] and Farhad Rachidi [3]

[1] Department of Electrical Engineering, Bu-Ali Sina University, 65178 Hamedan, Iran;
    sajadmohamadi.el@gmail.com
[2] University of Applied Sciences of Western Switzerland (HES-SO), 1400 Yverdon-les-Bains, Switzerland;
    mohammad.azadifar@heig-vd.ch
[3] Electromagnetic Compatibility Laboratory, Swiss Federal Institute of Technology (EPFL), 1015 Lausanne,
    Switzerland; farhad.rachidi@epfl.ch
* Correspondence: hamidr.karami@basu.ac.ir; Tel.: +98-8292505-216

Received: 4 March 2020; Accepted: 23 March 2020; Published: 30 March 2020

Abstract: An open accelerator (OpenACC)-aided graphics processing unit (GPU)-based finite difference time domain (FDTD) method is presented for the first time for the 3D evaluation of lightning radiated electromagnetic fields along a complex terrain with arbitrary topography. The OpenACC directive-based programming model is used to enhance the computational performance, and the results are compared with those obtained by using a CPU-based model. It is shown that OpenACC GPUs can provide very accurate results, and they are more than 20 times faster than CPUs. The presented results support the use of OpenACC not only in relation to lightning electromagnetics problems, but also to large-scale realistic electromagnetic compatibility (EMC) applications in which computation time efficiency is a critical factor.

Keywords: graphics processing unit (GPU); OpenACC (open accelerators); finite difference time domain (FDTD); lightning magnetic fields

---

1. Introduction

Rigorous modeling and solution of large-scale electromagnetic compatibility (EMC) problems often require prohibitive computational resources. Fast algorithms and techniques, as well as hardware platforms with high pipelining capability, are usually used to circumvent this problem (e.g., [1–15]). Graphics processing units (GPUs) are characterized by a high parallelism capability. Figure 1 presents schematically the CPU and GPU architectures. As can be seen in Figure 1, the number of threads in GPUs is dramatically higher than that in CPUs. A CPU consists of a few immense arithmetic logic unit (ALU) cores aiming to process with high cache memory and one enormous control module capable of managing a few threads at the same time. On the other hand, a GPU includes many small ALUs, small control modules, and a small cache. Furthermore, it can execute thousands of threads concurrently since it is optimized for parallel operations [16]. One of the particular examples of a large-scale EMC problem is the evaluation of lightning electromagnetic fields, and coupling to structures. It is a large-scale problem because of (i) the scale of the solution domain, which can be in the order of several tens of kilometers, and (ii) the complexity of the propagation media, when considering the inhomogeneity and roughness of the soil (e.g., [17–24]). In most of the studies focused on the evaluation of lightning radiated electromagnetic fields and their induced disturbances on nearby structures, such as overhead transmission lines and buried cables (e.g., [25–31]), computations have been carried out by making use of CPU-based systems.

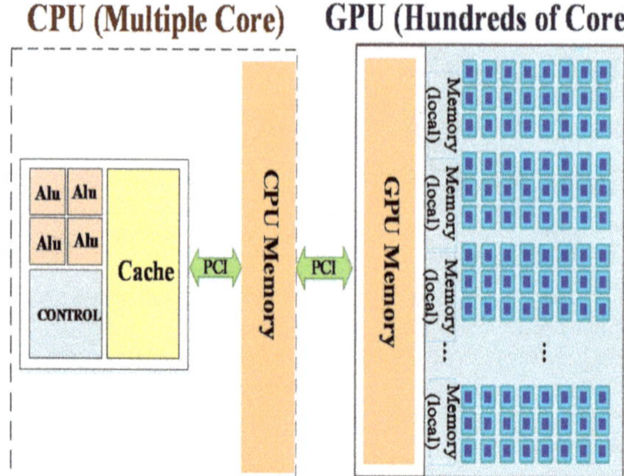

Figure 1. CPU and graphics processing unit (GPU) architectures.

However, there exist several attempts in which different hardware platforms have been used for the same study (e.g., [32–36]). One example is the use of GPU-based compute unified device architecture (CUDA) programming for the finite difference time domain (FDTD) evaluation of lightning electromagnetic fields. Due to its high-speed calculation, this approach facilitates three-dimensional modeling of a problem, taking into account parameter uncertainties such as irregular lightning channel and surface roughness [34]. It is noted that this approach is supposedly 100 times faster than the serial processing approach in CPU [33]. Although the GPU-based CUDA programming approach is highly efficient and provides the programmer with the flexibility to utilize various memories such as cache memory, it requires the involvement of the programmer for the determination of many programming details [37]. The OpenACC programming model suggested by NVIDIA, CAPS, Cray, and the Portland Group is a general user-driven directive-based parallel programming model developed for engineers and scientists in OpenACC. The programmer has the capability of incorporating compiler directives and library routines into FORTRAN, C, or C++ source codes to assign the area within the code that needs expedition in parallel on GPU. In fact, the programmer is not preoccupied with parallelism details and, in turn, leaves these tasks to the compiler. This programming model efficiently and intelligently launches the kernels and parallels the code onto the GPU. OpenACC was proposed for the first time in 2011 as a high-level programming approach that has offered almost all types of processors with high performance and portability. This programming model allows executing and creating codes using the available and future graphics hardware [38]. This is because OpenACC can transfer calculations and data from the host to the accelerator device. Despite their different architectures, the two former hardware components (host and accelerator device) and their corresponding memories are either shared or separated. An accelerated computing model of OpenACC is presented in Figure 2. According to Figure 2, the OpenACC compiler executes the code and manages the transferred data between the host and accelerator device. Furthermore, OpenACC benefits from a top-level compiler directive for the sake of portability to parallelize different sections of the code and use parallel and optimized compilers to develop and execute the corresponding codes. Several compilation directives or programs can be defined by OpenACC for parallel execution of code fragments. Recently, graphical processing using OpenACC has caught the attention of many researchers due to its relative simplicity and high performance (e.g., [39,40]). One of the particular examples of using OpenACC on graphics processors is to calculate the vector potential using the finite difference generated by time-domain Green's functions of layered media where the computational speed was 45.97 times the CPU computational speed on MATLAB [40].

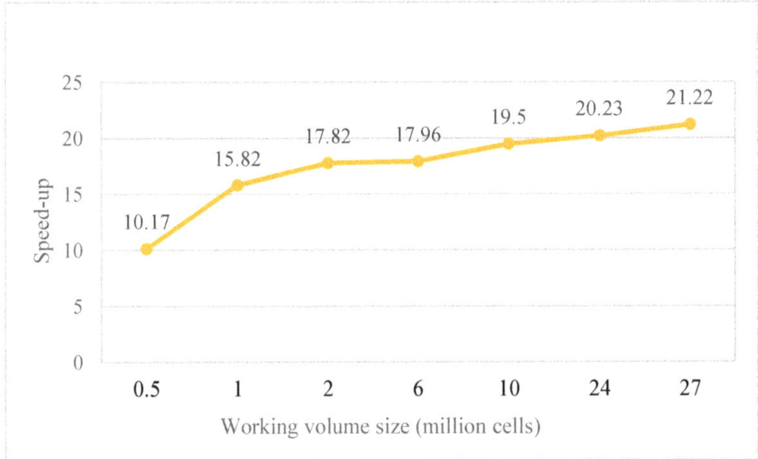

Figure 2. Open accelerator (OpenACC) model [38].

In this paper, OpenACC-based GPU processing is used to tackle the problem of long computation time in the FDTD method for the evaluation of electromagnetic fields generated by a lightning channel. It is worth noting that OpenACC benefits from relatively simple programming and dramatically increases computational speed. The rest of the paper is organized as follows. Section 2 presents the required steps for executing the computational FDTD code in graphics processing using OpenACC. Section 3 describes the adopted models and computational methods, as well as FDTD parameters. Section 4 presents the processing speed for the proposed method, and the results are compared with CPU serial processing speed. Concluding remarks are provided in Section 5.

2. OpenACC Implementation

2.1. FDTD Algorithm

The flowchart of the FDTD algorithm to calculate lightning electromagnetic fields consists of five main modules [41,42], as shown in Figure 3. In the FDTD algorithm, first, all the variables and parameters are set and defined in the initialization step, followed by the assignment of the required memory to store each component of the electromagnetic field. In each time step, the lightning channel source model excites the desired space. Then, the components of the electric (E_x, E_y, E_z) and magnetic (H_x, H_y, H_z) fields are computed and updated. Finally, boundary conditions are applied to avoid unwanted reflections of the electromagnetic field. These steps are repeated until the simulations are performed in the desired time period. The output of the FDTD algorithm would be the electric and magnetic fields at each observation point within the specified time period.

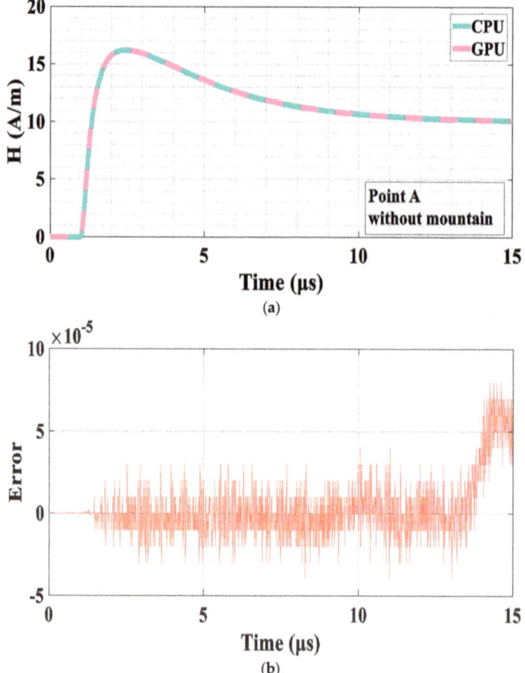

Figure 3. Finite difference time domain (FDTD) algorithm flowchart.

2.2. Hardware and Software Used

For the GPU implementation of the adopted FDTD approach, the GeForce GTX 1050 Ti hardware and PGI Community Edition 17.10 software were used. The PGI compiler was developed by the Portland Group and NVidias [43], and the PGI Community Edition includes a free license of the PGI C++ compilers and tools for multicore CPUs and NVIDIA Tesla GPUs. Community Edition enables the development of performance-portable High-performance computing (HPC) applications with a uniform source code in the most widely used parallel processors and systems [44]. Details of the GPU are presented in Table 1. For comparison purposes, we also repeated the calculations on CPU by making use of an Intel® Core™ i7-6800K @ 3.40 GHz hardware.

Table 1. GPU specifications.

Name	GeForce GTX 1050 Ti
GPU Architecture	Pascal
NVIDIA CUDA® Cores	760
Standard Memory	4 GB GDDR5
Memory Speed	7 Gbps
Graphics Clock (MHz)	1290
Processor Clock (MHz)	1392

2.3. OpenACC Programming

This section aims to provide more elaborate details about OpenACC programming, as well as compiler specifications for building and executing codes on the GPU.

2.3.1. OpenACC Data Clause

The data required for the process implementation are transferred to the GPU memory prior to the parallel region in the code, and the appropriate OpenACC directives are then inserted into the main code for each loop. Since it is burdensome for a compiler to determine whether data will be needed in the future, they are copied conscientiously in case of future demand. Data locality helps to solve this problem so that the data remain in the local memory in the host or system as long as needed. Data clauses enable the programmer to control the method and time of generating/copying the data from/on the device. The data directives are briefly described as follows:

- Copy: Provision of space for the variables listed on the device, initialization of the variables via copying data on the device at the region's first parts, copying the generated results on the host at the region's last parts, and, finally, releasing the space on the device [38,45];
- Copyin: Provide the required space for the variables available on the device list, carry out the initialization of the variables via copying data on the device at the region's beginning, and free the space on the device without copying the data back onto the host [38,45].

In the FDTD calculations, the start and end points of data locality were positioned before the time loop (iteration) and at the end of all modules and functions, respectively. As shown in Figure 4, the starting point of data locality in the FDTD algorithm was placed before the time loop (iteration) in the code so that the proper data needed for the process execution were moved to the GPU memory using copy directive. The ending point of data locality was also placed after performing (execution) all of the modules and functions so that the generated output data returned to the CPU memory using copyin directive. This data directive avoids extra and inefficient data transfer for the calculation. Therefore, the data transfer time between the GPU and CPU is reduced using the data clause, remarkably reducing the computation time.

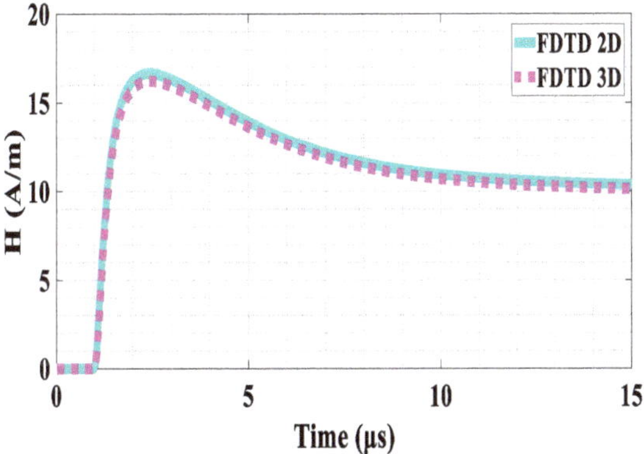

Figure 4. Optimized data locality.

2.3.2. OpenACC Parallelize Loops

Parallel and kernel regions are two different approaches to incorporate parallelism into the code provided by OpenACC:

- Structure of kernels: The structure of kernels determines a code region that can contain parallelism. However, the automatic parallelization capabilities of the compiler, identification of the loops that are secure enough for parallelization, and acceleration of these loops influence the analysis of the region. The compiler is free to select the best way of mapping the parallelism in the loops onto the hardware [38,45];

- Structure of parallel: If this directive is put in a loop, the programmer claims that the affected parallelization loop is secured and enables the compiler to choose how to schedule the loop iterations on the target accelerator. In this case, the programmer determines the availability of parallelism per se, whereas the decision regarding mapping parallelism onto the accelerator depends on the compiler's knowledge of the device [38,45].

The OpenACC parallelizing directives were applied to different parts of the code to implement parallel computing. For example, as can be seen in Figure 5, the #pragma ACC kernels loop is set at the beginning of each of the three loops, which calculate and update each of the electromagnetic components. Table 2 presents the computation time associated with these two directives. According to Table 2, C on CPU runtime (without optimization) is 2231.59 s, which is the benchmark time. The speed was increased by a factor of 2.43 by optimizing the programming C on CPU. Moreover, as can be observed in the table, the computing speed on the GPU processor increased by factors of 10.49 and 11.14 by using different directive OpenACC including the #pragma ACC parallel loop and the #pragma ACC kernels loop OpenACC, respectively.

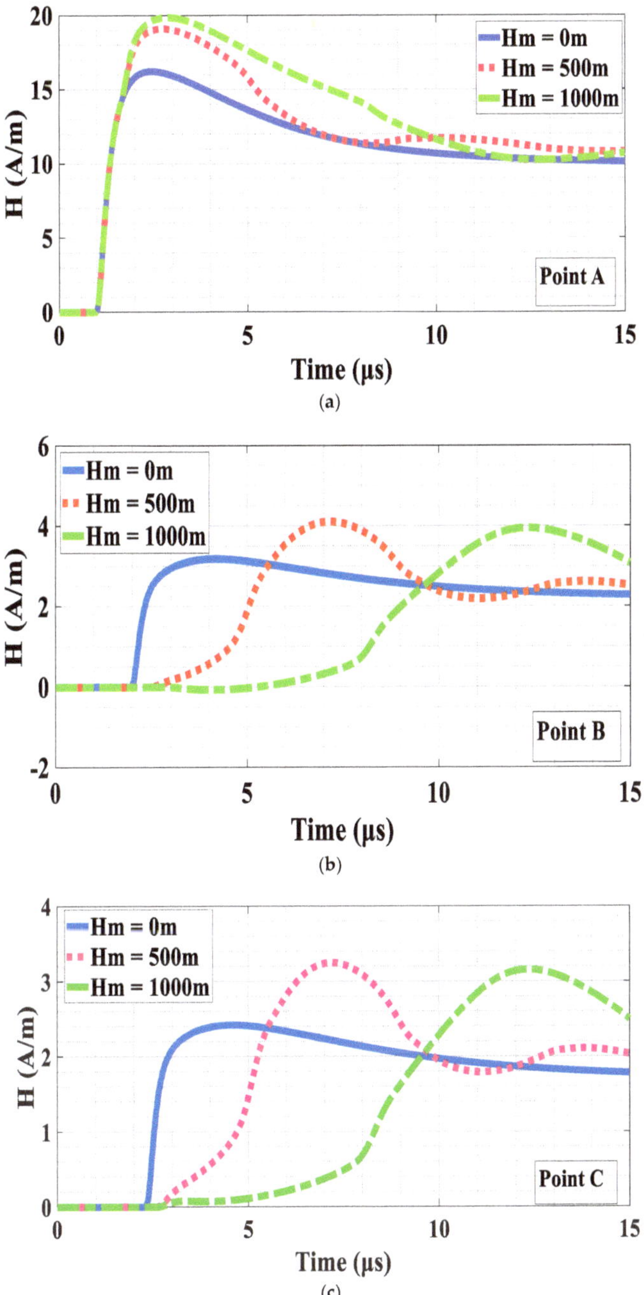

Figure 5. Mapping parallelism onto the hardware.

Table 2. Comparison of OpenACC directive.

Programming Language	Directive	Runtime(s)	Speed-Up
C (without optimization)	-	2231.59	1 X
C (with optimization)	-	914.98	2.43 X
OpenACC	#pragma ACC parallel loop #pragma ACC kernels loop	212.55 200.21	10.49 X

2.3.3. Save Variable

As shown in Figure 1, a GPU processor is presented as a set of multiprocessors, each with its own stream processors and memory. The stream processors have the full capability to execute integer and single precision floating point arithmetic, including extra cores applied to double-precision procedure [16]. The results of Table 3 reveal that when the variables are of the float type, the calculation time in the GPU process is reduced. Precisely, in GPU processing by storing variables with single precision, the computing speed can be increased up to 20.02 times the speed of storing variables by double processing in the CPU series.

Table 3. Increase in the computational speed regarding the accuracy of the variables stored.

Programming Language	Variable Precision	Runtime(s)	Speed-Up
C (without optimization)	double float	2231.59 2269.56	1 X 0.98 X
OpenACC	double float	200.51 111.47	11.01 X 20.02 X

2.3.4. Optimization by Tuning Number of Vectors

The OpenACC performance model includes three sections: Gang (Thread block), Worker (Warp), and Vector (Threads in a Warp) [38].

Gangs, vectors, and workers can be automatically set up by the OpenACC compiler or by the developer to yield the optimum approach. The number of vectors is modified in the OpenACC directives to gain the maximum occupancy and speed-up in GPU. Modifying the vector length clause can be beneficial to reaching the optimal runtime value by trying different vector lengths for the accelerator, which was used in this study. Figure 6 shows the relative runtime derived by varying the vector length compared to the compiler-selected value. It is worth noting that the best performance was obtained for a vector length of 1024.

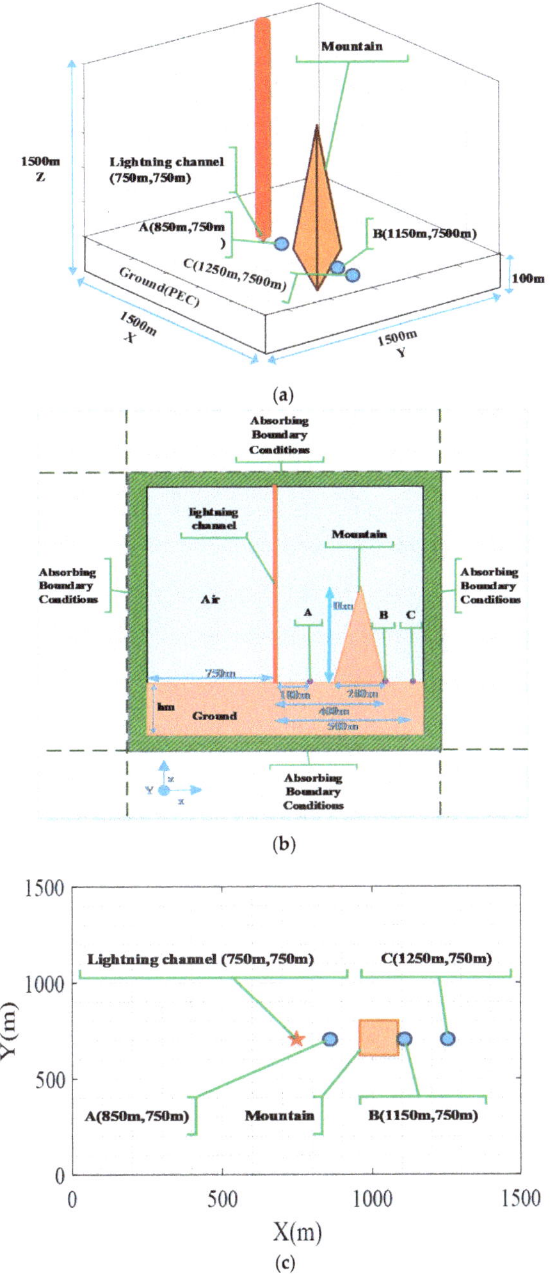

Figure 6. Relative runtime from varying vector length from the default value.

3. Application: Models and Hypotheses Considered for Analysis

The geometry of the problem is shown in Figure 7. The aim is to evaluate lightning electromagnetic fields over a mountainous terrain by making use of the FDTD technique. We assume a pyramidal mountain with a height H_m, which is varied from 0 (flat ground case) to 1000 m, while its cross-section is assumed to have an area of A = 40,000 m^2. The ground and mountain are assumed to be a perfect electric conductor (PEC) and the ground depth h$_m$ = 100 m. The vertical electric field and the radial magnetic field components are calculated for three observation points A, B, and C. These points are, respectively, 100, 400, and 500 m far from the lightning channel base, and all located on the ground surface. In the FDTD calculations, a vertical array of current sources was used for the modeling of the lightning channel [46]. The modified transmission line model with exponential decay (MTLE) [47,48] was used for modeling the return stroke channel. For subsequent stroke current waveforms, a summation of two Heidler functions was used [49], which is given in Equation (1):

$$i(t) = \left[\frac{I_{01}}{\eta_1} \frac{\left(\frac{t}{\tau_{11}}\right)^{n_1}}{1+\left(\frac{t}{\tau_{11}}\right)^{n_1}} e^{-\frac{t}{\tau_{21}}} + \frac{I_{02}}{\eta_2} \frac{\left(\frac{t}{\tau_{12}}\right)^{n_2}}{1+\left(\frac{t}{\tau_{12}}\right)^{n_2}} e^{-\frac{t}{\tau_{22}}} \right] \tag{1}$$

where I_{01} = 10.7 kA, τ_{11} = 0.25 μs, τ_{21} = 2.5 μs, n_1 = 2, η_1 = 0.639, I_{01} = 6.5 kA, τ_{11} = 0.25 μs, τ_{21} = 230 μs, n_1 = 2, and η_1 = 0.867, respectively [49]. The current decay constant and the return stroke speed are considered to be λ = 2000 m and v = 1.5 × 10^8 m/s, respectively. The time increment was set to 9.5 ns. The dimensions of the computational domain are 1500 × 1500 × 1500 m with a spatial mesh grid composed of square cells of 5 × 5 × 5 m. In an unbounded space, the electromagnetic response analysis of a structure absorbing boundary conditions in which unwanted reflections are suppressed must be applied to planes to truncate and limit the open space and the working volume, respectively. In order to avoid reflections at the domain boundaries, Liao's condition is adopted as the absorption boundary condition [50]. The horizontal magnetic field (H$_y$) components at the observation points A, B, and C are calculated assuming different heights for the mountain. The results are presented in Figure 8.

According to Figure 8, the time delay, peak value, waveshape, and amplitudes of the horizontal magnetic fields (H$_y$) are affected by the presence of the mountain. Although the effect of the mountain on the time delay (peak field and onset time) was negligible for point A, its effect on points B and C (right side of the hill) was significant [51]. The results represented in Figure 8 have been validated using canonical structures against the simulation results obtained in [51]. The only difference was that the GPU was used and the dimensions were 1500 × 1500 × 1500 m in this study, whereas a CPU processor was used in [51]. Figure 9 presents, for validation purposes, the obtained results compared with those obtained using a 2D-FDTD method.

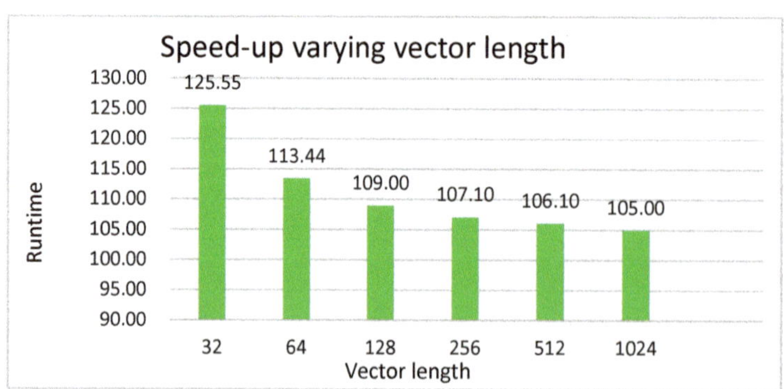

Figure 7. The simulation domain of the 3D-FDTD method and geometry for the lightning electromagnetic fields over mountainous terrain: (**a**) three-dimensional perspective, (**b**) side view, (**c**) top view.

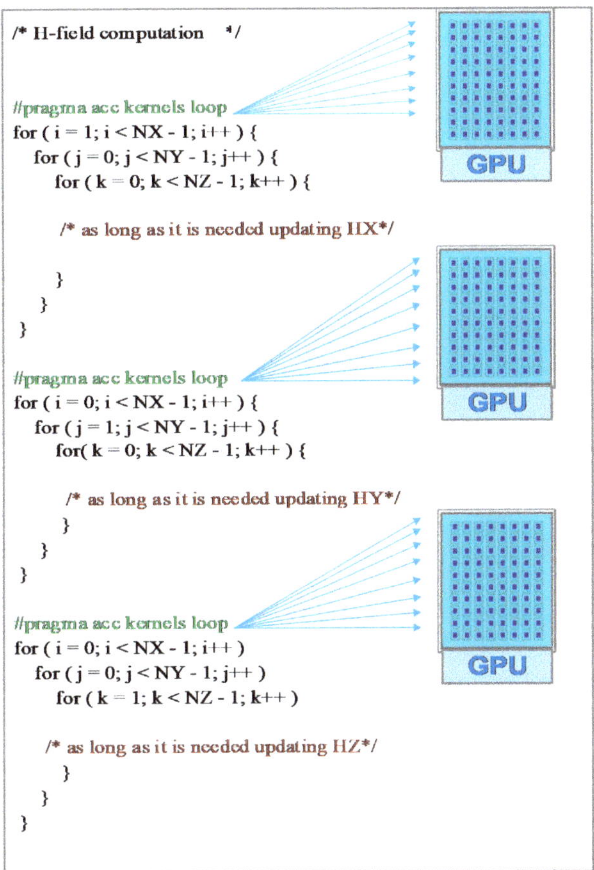

Figure 8. Horizontal magnetic field (H_y) associated with a subsequent return stroke at the considered 3 observation points A (**a**), B (**b**), C (**c**), and considering different mountain heights (H_m = 0, 200, 500, and 1000 m).

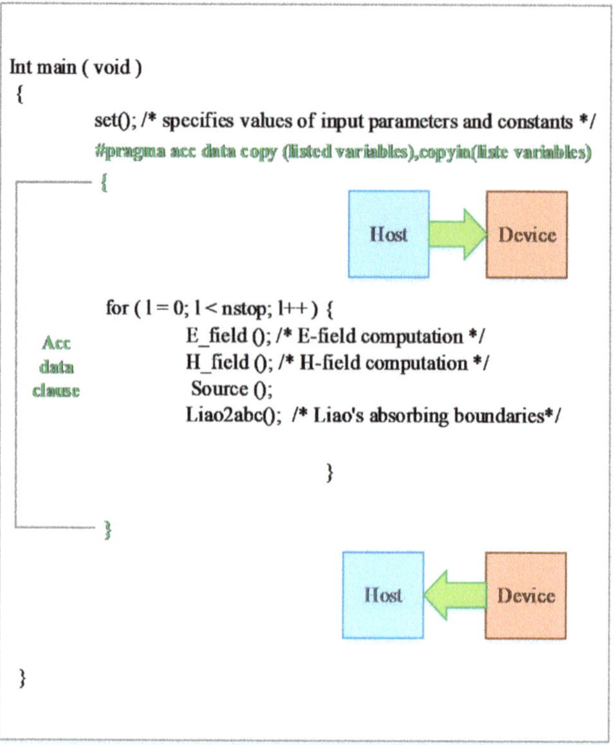

Figure 9. Horizontal magnetic field (H_y) on a smooth ideal ground at observation point A. The height of the mountain is set at $H_m = 0$ m. Results were calculated using the 2D-FDTD (solid line) and the 3D-FDTD (dashed line) methods.

4. Performance Analysis

This section compares the graphics processing time using OpenACC for the considered 3D-FDTD problem in this study (dimensions of the volume $1500 \times 1500 \times 1500$ m), with the execution time of this algorithm using a serial C programming language in CPU. Table 4 presents the comparison of the performance of OpenACC compared to CPU serial C. As can be seen in Table 4, the OpenACC execution time in an efficient state was 21.22 times that of the serial processing in the CPU without compromising precision. As can be seen in Table 3, the calculation time in the GPU process was drastically reduced by storing the variables of the float type (single precision). Although float variables (single precision) led to a decrease in the accuracy of the computation, this reduction was negligible. Figure 10 depicts the difference (computing error) between CPU and GPU processing for point A, at which the mountain height was zero ($H_m = 0$). The error is smaller than 0.008% and can be considered insignificant. In addition, Figure 11 indicates the gain in the computation speed as a function of the size of the workspace. It can be seen that the GPU performance increases with larger simulation domains.

Table 4. Performance of OpenACC implementation.

Programming Language	Processing Time(s)	Speed-Up
CPU serial C	2231.59	1 X
OpenAcc	105.15	21.22 X

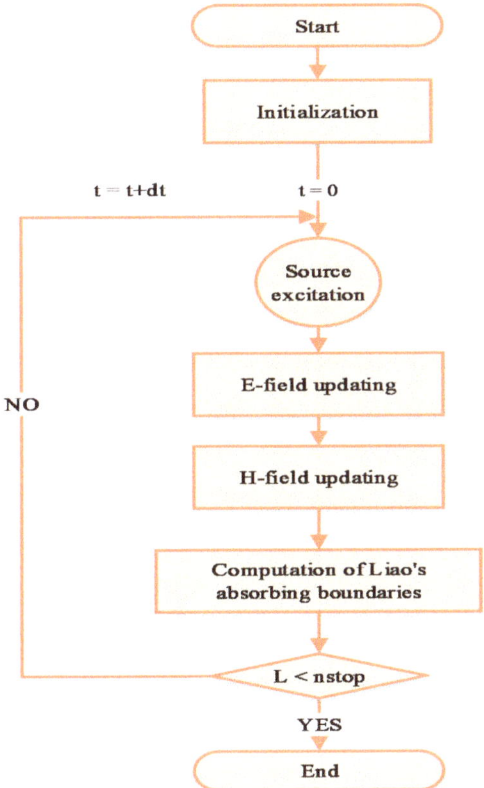

Figure 10. (**a**) The blue solid line and pink line represent the CPU and GPU processing, respectively, (**b**) Numerical error between the CPU and GPU processing.

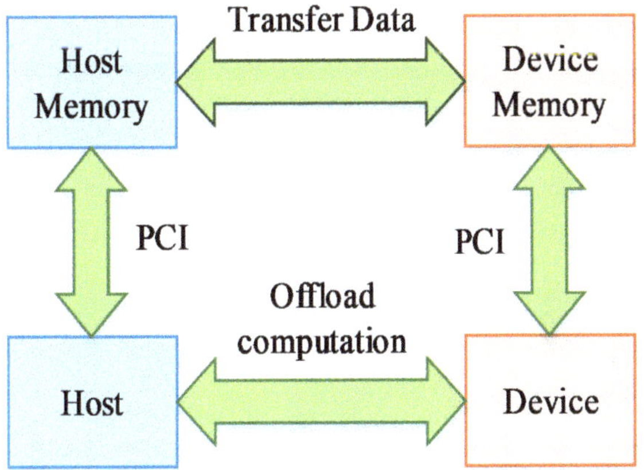

Figure 11. Gain in GPU runtime with respect to CPU for various numbers of nodes.

5. Conclusions

An OpenACC-aided graphics processing unit (GPU)-based finite difference time domain (FDTD) method was presented for the first time for the 3D evaluation of lightning radiated electromagnetic fields along a complex terrain with arbitrary topography. The OpenACC (Open accelerators) directive-based programming model was used to enhance the computational performance and the results were compared with those obtained by using a CPU-based model. It was shown that OpenACC GPUs can provide very accurate results while being more than 20 times faster than CPUs. The presented results support the use of OpenACC not only for lightning electromagnetics problems, but also for large-scale realistic EMC applications in which computation time efficiency is a critical factor.

Author Contributions: Conceptualization, H.K.; methodology, S.M and H.K.; software, S.M; validation, S.M; resources, H.K.; writing—original draft preparation, S.M.; writing—review and editing, H.K., M.A., and F.R.; supervision, H.K. All authors have read and agreed to the published version of the manuscript.

Funding: This work was supported by the Bakhtar Regional Electricity Company, Arak, Iran.

Conflicts of Interest: The authors declare no conflict of interest.

References

1. Balevic, A.; Rockstroh, L.; Tausendfreund, A.; Patzelt, S.; Goch, G.; Simon, S. Accelerating simulations of light scattering based on finite-difference time-domain method with general purpose GPUs. In Proceedings of the 2008 11th IEEE International Conference on Computational Science and Engineering, Sao Paulo, Brazil, 16–18 July 2008; IEEE: Piscataway, NJ, USA; pp. 327–334.

2. Bracken, E.; Polstyanko, S.; Lambert, N.; Suda, R.; Giannacopulos, D.D. Power aware parallel 3-D finite element mesh refinement performance modeling and analysis with CUDA/MPI on GPU and multi-core architecture. *IEEE Trans. Magn.* **2012**, *48*, 335–338. [CrossRef]

3. Cannon, P.D.; Honary, F. A GPU-accelerated finite-difference time-domain scheme for electromagnetic wave interaction with plasma. *IEEE Trans. Antennas Propag.* **2015**, *63*, 3042–3054. [CrossRef]

4. Chen, Y.; Zuo, S.; Zhang, Y.; Zhao, X.; Zhang, H. Large-scale parallel method of moments on CPU/MIC heterogeneous clusters. *IEEE Trans. Antennas Propag.* **2017**, *65*, 3782–3787. [CrossRef]

5. Fan, T.-Q.; Guo, L.-X.; Liu, W. A novel OpenGL-based MoM/SBR hybrid method for radiation pattern analysis of an antenna above an electrically large complicated platform. *IEEE Trans. Antennas Propag.* **2015**, *64*, 201–209. [CrossRef]

6. Guan, J.; Yan, S.; Jin, J.-M. An accurate and efficient finite element-boundary integral method with GPU acceleration for 3-D electromagnetic analysis. *IEEE Trans. Antennas Propag.* **2014**, *62*, 6325–6336. [CrossRef]

7. Ikuno, S.; Fujita, Y.; Hirokawa, Y.; Itoh, T.; Nakata, S.; Kamitani, A. Large-scale simulation of electromagnetic wave propagation using meshless time domain method with parallel processing. *IEEE Trans. Magn.* **2013**, *49*, 1613–1616. [CrossRef]

8. Kaburcuk, F.; Demir, V.; Elsherbeni, A.Z.; Arvas, E.; Mautz, J.R. Time-Domain Iterative Multiregion Technique for 3-D Scattering and Radiation Problems. *IEEE Trans. Antennas Propag.* **2016**, *64*, 1807–1817. [CrossRef]

9. Masumnia-Bisheh, K.; Forooraghi, K.; Ghaffari-Miab, M.; Furse, C.M. Geometrically Stochastic FDTD Method for Uncertainty Quantification of EM Fields and SAR in Biological Tissues. *IEEE Trans. Antennas Propag.* **2019**, *67*, 7466–7475. [CrossRef]

10. Park, S.-M.; Kim, E.-K.; Park, Y.B.; Ju, S.; Jung, K.-Y. Parallel dispersive FDTD method based on the quadratic complex rational function. *IEEE Antennas Wirel. Propag. Lett.* **2015**, *15*, 425–428. [CrossRef]

11. Sypek, P.; Dziekonski, A.; Mrozowski, M. How to render FDTD computations more effective using a graphics accelerator. *IEEE Trans. Magn.* **2009**, *45*, 1324–1327. [CrossRef]

12. Unno, M.; Aono, S.; Asai, H. GPU-based massively parallel 3-D HIE-FDTD method for high-speed electromagnetic field simulation. *IEEE Trans. Electromagn. Compat.* **2012**, *54*, 912–921. [CrossRef]

13. Wan, J.; Lei, J.; Liang, C.-H. An efficient analysis of large-scale periodic microstrip antenna arrays using the characteristic basis function method. *Prog. Electromagn. Res.* **2005**, *50*, 61–81. [CrossRef]

14. Xu, K.; Fan, Z.; Ding, D.-Z.; Chen, R.-S. GPU accelerated unconditionally stable Crank-Nicolson FDTD method for the analysis of three-dimensional microwave circuits. *Prog. Electromagn. Res.* **2010**, *102*, 381–395. [CrossRef]

15. Zainud-Deen, S.H.; Hassan, E.; Ibrahim, M.S.; Awadalla, K.H.; Botros, A.Z. Electromagnetic scattering using GPU-based finite difference frequency domain method. *Prog. Electromagn. Res.* **2009**, *16*, 351–369. [CrossRef]

16. Micikevicius, P. 3D finite difference computation on GPUs using CUDA. In Proceedings of the 2nd Workshop on General Purpose Processing on Graphics Processing Units, Washington, DC, USA, 8 March 2009; pp. 79–84.

17. Karami, H.; Rachidi, F.; Rubinstein, M. On practical implementation of electromagnetic models of lightning return-strokes. *Atmosphere* **2016**, *7*, 135. [CrossRef]

18. Li, D.; Azadifar, M.; Rachidi, F.; Rubinstein, M.; Paolone, M.; Pavanello, D.; Metz, S.; Zhang, Q.; Wang, Z. On lightning electromagnetic field propagation along an irregular terrain. *IEEE Trans. Electromagn. Compat.* **2015**, *58*, 161–171. [CrossRef]

19. Li, D.; Luque, A.; Rachidi, F.; Rubinstein, M. Evaluation of Ionospheric D region Parameter Using Simultaneously Measured Lightning Current and 380-km Distant Electric Field. In Proceedings of the AGU Fall Meeting Abstracts, Washington, DC, USA, 10–14 December 2018.

20. Li, D.; Rachidi, F.; Rubinstein, M. FDTD Modeling of lightning electromagnetic field propagation over mountainous terrain. In Proceedings of the 2019 International Applied Computational Electromagnetics Society Symposium (ACES), Miami, FL, USA, 14–19 April 2019; pp. 1–2.

21. Li, D.; Rubinstein, M.; Rachidi, F.; Diendorfer, G.; Schulz, W.; Lu, G. Location accuracy evaluation of ToA-based lightning location systems over mountainous terrain. *J. Geophys. Res. Atmos.* **2017**, *122*, 11760–11775. [CrossRef]

22. Mostajabi, A.; Li, D.; Azadifar, M.; Rachidi, F.; Rubinstein, M.; Diendorfer, G.; Schulz, W.; Pichler, H.; Rakov, V.A.; Pavanello, D. Analysis of a bipolar upward lightning flash based on simultaneous records of currents and 380-km distant electric fields. *Electr. Power Syst. Res.* **2019**, *174*, 105845. [CrossRef]

23. Oikawa, T.; Sonoda, J.; Sato, M.; Honma, N.; Ikegawa, Y. Analysis of lightning electromagnetic field on large-scale terrain model using three-dimensional MW-FDTD parallel computation. *IJTFM* **2012**, *132*, 44–50. [CrossRef]

24. Oikawa, T.; Sonoda, J.; Honma, N.; Sato, M. Analysis of Lighting Electromagnetic Field on Numerical Terrain and Urban Model using Three-Dimensional MW-FDTD Parallel Computation. *Electron. Commun. Japan* **2017**, *100*, 76–82. [CrossRef]

25. Paknahad, J.; Sheshyekani, K.; Rachidi, F. Lightning electromagnetic fields and their induced currents on buried cables. Part I: The effect of an ocean–land mixed propagation path. *IEEE Trans. Electromagn. Compat.* **2014**, *56*, 1137–1145. [CrossRef]

26. Paknahad, J.; Sheshyekani, K.; Rachidi, F.; Paolone, M. Lightning electromagnetic fields and their induced currents on buried 999bles. Part II: The effect of a horizontally stratified ground. *IEEE Trans. Electromagn. Compat.* **2014**, *56*, 1146–1154. [CrossRef]

27. Paknahad, J.; Sheshyekani, K.; Rachidi, F.; Paolone, M.; Mimouni, A. Evaluation of lightning-induced currents on cables buried in a lossy dispersive ground. *IEEE Trans. Electromagn. Compat.* **2014**, *56*, 1522–1529. [CrossRef]

28. Tatematsu, A.; Rachidi, F.; Rubinstein, M. Analysis of electromagnetic fields inside a reinforced concrete building with layered reinforcing bar due to direct and indirect lightning strikes using the FDTD method. *IEEE Trans. Electromagn. Comput.* **2015**, *57*, 405–417. [CrossRef]

29. Tatematsu, A.; Rachidi, F.; Rubinstein, M. A technique for calculating voltages induced on twisted-wire pairs using the FDTD method. *IEEE Trans. Electromagn. Compat.* **2016**, *59*, 301–304. [CrossRef]

30. Karami, H.; Sheshyekani, K.; Rachidi, F. Mixed-potential integral equation for full-wave modeling of grounding systems buried in a lossy multilayer stratified ground. *IEEE Trans. Electromagn. Compat.* **2017**, *59*, 1505–1513. [CrossRef]

31. Šunjerga, A.; Gazzana, D.S.; Poljak, D.; Karami, H.; Sheshyekani, K.; Rubinstein, M.; Rachidi, F. Tower and path-dependent voltage effects on the measurement of grounding impedance for lightning studies. *IEEE Trans. Electromagn. Compat.* **2018**, *61*, 409–418. [CrossRef]

32. Tatematsu, A.; Noda, T. Three-dimensional FDTD calculation of lightning-induced voltages on a multiphase distribution line with the lightning arresters and an overhead shielding wire. *IEEE Trans. Electromagn. Compat.* **2013**, *56*, 159–167. [CrossRef]

33. Pyrialakos, G.; Zygiridis, T.; Kantartzis, N.; Tsiboukis, T. FDTD analysis of 3D lightning problems with material uncertainties on GPU architecture. In Proceedings of the 2014 International Symposium on Electromagnetic Compatibility, Gothenburg, Sweden, 1–4 September 2014; pp. 577–582.

34. Pyrialakos, G.; Zygiridis, T.; Kantartzis, N.; Tsiboukis, T. GPU-based three-dimensional calculation of lightning-generated electromagnetic fields. In Proceedings of the 2014 International Conference on Numerical Electromagnetic Modeling and Optimization for RF, Microwave, and Terahertz Applications (NEMO), Pavia, Italy, 14–16 May 2014; pp. 1–4.

35. Pyrialakos, G.G.; Zygiridis, T.T.; Kantartzis, N.V.; Tsiboukis, T.D. GPU-based calculation of lightning-generated electromagnetic fields in 3-D problems with statistically defined uncertainties. *IEEE Trans. Electromagn. Compat.* **2015**, *57*, 1556–1567. [CrossRef]

36. Tatematsu, A. Overview of the three-dimensional FDTD-based surge simulation code VSTL REV. In Proceedings of the 2016 Asia-Pacific International Symposium on Electromagnetic Compatibility (APEMC), Shenzhen, China, 17–21 May 2016; pp. 670–672.

37. Al-Jarro, A.; Clo, A.; Bagci, H. Implementation of an explicit time domain volume integral equation solver on gpus using openacc. In Proceedings of the 29th International Review of Progress in Applied Computational Electromagnetics, Monterey, CA, USA, 24–28 March 2013.

38. OpenACC Programming and Best Practices Guide. OpenACC.org, June 2015. Available online: https://www.openacc.org/sites/default/files/inline-files/OpenACC_Programming_Guide_0.pdf (accessed on 8 November 2019).

39. Rostami, S.R.M.; Ghaffari-Miab, M. Fast computation of finite difference generated time-domain green's functions of layered media using OpenAcc on graphics processors. In Proceedings of the 2017 Iranian Conference on Electrical Engineering (ICEE), Tehran, Iran, 2–4 May 2017; pp. 1596–1599.

40. Rostami, S.R.M.; Ghaffari-Miab, M. Finite difference generated transient potentials of open-layered media by parallel computing using OPENMP, MPI, OPENACC, and CUDA. *IEEE Trans. Antennas Propag.* **2019**, *67*, 6541–6550. [CrossRef]

41. Gedney, S.D. Introduction to the finite-difference time-domain (FDTD) method for electromagnetics. *Synth. Lect. Comput. Electromagn.* **2011**, *6*, 1–250. [CrossRef]

42. Kunz, K.S.; Luebbers, R.J. *The Finite Difference Time Domain Method for Electromagnetics*; CRC press: Boca Raton, FL, USA, 1993.

43. The Portland Group—PGI. Available online: https://www.pgroup.com/about/ (accessed on 25 May 2018).

44. PGI Community Edition. Available online: https://www.pgroup.com/products/community.htm (accessed on 25 May 2018).

45. Chandrasekaran, S.; Juckeland, G. *OpenACC for Programmers: Concepts and Strategies*; Addison-Wesley Professional: Boston, MA, USA, 2017.

46. Li, D.; Zhang, Q.; Wang, Z.; Liu, T. Computation of lightning horizontal field over the two-dimensional rough ground by using the three-dimensional FDTD. *IEEE Trans. Electromagn. Compat.* **2013**, *56*, 143–148. [CrossRef]

47. Nucci, C.A. On lightning return stroke models for LEMP calculations. In Proceedings of the 19th International Conference Lightning Protection, Graz, Austria, 25–29 April 1988.

48. Rachidi, F.; Nucci, C. On the Master, Uman, Lin, Standler and the modified transmission line lightning return stroke current models. *J. Geophys. Res. Atmos.* **1990**, *95*, 20389–20393. [CrossRef]

49. Rachidi, F.; Janischewskyj, W.; Hussein, A.M.; Nucci, C.A.; Guerrieri, S.; Kordi, B.; Chang, J.-S. Current and electromagnetic field associated with lightning-return strokes to tall towers. *IEEE Trans. Electromagn. Compat.* **2001**, *43*, 356–367. [CrossRef]

50. Baba, Y.; Rakov, V.A. *Electromagnetic Computation Methods for Lightning Surge Protection Studies*; John Wiley & Sons: Hoboken, NJ, USA, 2016. [CrossRef]

51. Li, D.; Paknahad, J.; Rachidi, F.; Rubinstein, M.; Sheshyekani, K.; Zhang, Q.; Wang, Z. Propagation effects on lightning magnetic fields over hilly and mountainous terrain. In Proceedings of the 2015 IEEE International Symposium on Electromagnetic Compatibility (EMC), Dresden, Germany, 16–22 August 2015; pp. 1436–1440.

Article

Lightning Protection Methods for Wind Turbine Blades: An Alternative Approach

Viktor Mucsi [1], Ahmad Syahrir Ayub [1,*], Firdaus Muhammad-Sukki [1,*], Muhammad Zulkipli [2], Mohd Nabil Muhtazaruddin [3], Ahmad Shakir Mohd Saudi [4,*] and Jorge Alfredo Ardila-Rey [5]

[1] School of Engineering, Robert Gordon University, The Sir Ian Wood Building, Riverside East, Garthdee Road, AB10 7GJ Aberdeen, UK; v.mucsi@rgu.ac.uk

[2] Faculty of Engineering Technology, Universiti Tun Hussein Onn Malaysia, Pagoh Higher Education Hub, KM1, Jalan Panchor, Pagoh, Muar 84600 Johor, Malaysia; muhammad@uthm.edu.my

[3] Razak Faculty of Technology and Informatics, Universiti Teknologi Malaysia, Jalan Sultan Yahya Petra, Kuala Lumpur 54100, Malaysia; mohdnabil.kl@utm.my

[4] Universiti Kuala Lumpur-Institute of Medical Science Technology (UniKL MESTECH), A1-1, Jalan TKS 1, Taman Kajang Sentral, Kajang 43000, Selangor, Malaysia

[5] Department of Electrical Engineering, Universidad Técnica Federico Santa María, Santiago de Chile 8940000, Chile; jorge.ardila@usm.cl

* Correspondence: a.ayub2@rgu.ac.uk (A.S.A.); f.b.muhammad-sukki@rgu.ac.uk (F.M.-S.); ahmadshakir@unikl.edu.my (A.S.M.S.)

Received: 30 January 2020; Accepted: 10 March 2020; Published: 20 March 2020

Abstract: Lightning strikes happens in a fraction of time, where they can transfer huge amounts of charge and high currents in a single strike. The chances for a structure to be struck by lightning increases as the height increases; thus, tall structures are more prone to lightning. Despite the existing lightning protection systems available for wind turbine blades, there are still many cases reported due to the fact of damage caused by lightning strike. Owing to that, the present work introduces a new approach for a lightning protection system for wind turbine blades where preliminary investigations were done using Analysis Systems (ANSYS) Workbench. Two models were developed: one with a conventional type down conductor system and the other with a hybrid conductor system. The recorded findings have been compared and discussed, where it was found that the hybrid conductor system may provide alternative protection from lightning for wind turbine blades.

Keywords: lightning; lightning protection system; wind turbine blades; ANSYS workbench

---

1. Introduction

Windmills have been around for centuries, operating as grain grinders and water pumps. The concept and technology behind windmills have been adapted to generate electricity, which in its new form is now called wind turbines (i.e., wind energy). Wind energy generation is now becoming one of the largest contributors to renewable energy generation, where the recent demand for renewable energy has seen its increasing growth in use as well as in physical size. In other words, wind turbines are getting taller, in order to accommodate the demand, by capturing wind through a larger blade swept area and converting it into electricity. Owing to this, wind turbines are now more prone to lightning strikes due to the fact of their increased structural height.

There are approximately 2000 thunderstorms at any given minute and about 100 lightning strikes per second worldwide [1]. This creates great risk for tall structures, such as wind turbines, to be struck by lightning, where the average electric current from a lightning return stroke is 30 kA [1]. This massive flow of current can heat up the leader channel air to between 25,000 °C and 30,000 °C (around five times the effective temperature of the sun) [1–3]. Lightning protection system (LPS) is composed of lightning receptor, down conductor, and grounding, and all elements must be well connected to pass the lightning current to Earth safely. Although wind turbines are installed with LPS, there are still cases where blades and whole turbines are destroyed due to the fact of lightning strikes. Considering the 20–25 year design life for wind turbine [3], it is worth safeguarding the turbines from lightning strikes, because the damage associated with it will cause the down time of the turbine operation, causing extra costs for maintenance and an shortage of electricity. This, therefore, may suggest a need for improving the existing lightning protection systems for wind turbine blades.

2. Lightning Discharges and Existing Lightning Protection Systems for Wind Turbine

Mechanism of Lightning Discharge

Lightning discharge from cloud to ground stems from a stepped leader initiated in a cloud and increases the electric field within its path. When a grounded object is in that electric field, it generates a leader towards the stepped leader, and it is called a connecting leader. If the downward moving leader has a negative charge, then the connecting leader is positive. If the downward leader is negative, then the connecting leader is positive [3].

As a stepped leader approaches ground level or the tip of the grounded structures, the electric field increases to such an extent that it discharges, and connecting leaders starts to propagate towards the downward leader in an attempt to connect, to equal the potential difference. Taller structures generate longer connecting leaders due to the field enhancement caused by the accumulation of positive charge on the structure [1,3,4].

The stepped leader channel is at cloud potential, approximately 50 MV [1,3–8] and with the final connecting jump, a near of ground potential travels along the channel in the direction of the cloud, which is called return stroke. The flow of charge generates a large current with an average peak of 30 kA [1,3] to 80 kA [1]. Due to the rapid generation of heat of around 30,000 K [1,3] in the channel, a pressure is created of 10 atm or above [3]. In some instances, new charges from the cloud forming another electrical discharge called dart leaders, creating subsequent return strokes with an average peak current of about 10–15 kA [1,3].

Most negative cloud-to-ground flashes contain more than one stroke, generally 3–4 [1] and, in major cases, the first stroke is usually 2–3 [3] times larger than the following subsequent strokes. On the other hand, occasionally in multiple stroke flashes there is at least one subsequent stroke which is greater than the first return stroke [3].

3. Wind Turbine Blades and Its Protection Methods

3.1. Wind Turbines and Blades

There are two main types of wind turbines on the market nowadays: vertical axis and horizontal axis turbines. Due to the lower efficiency, vertical axis turbines were not considered in this paper. Modern turbines are dominantly composed of horizontal axis models, since with rotor blade pitching, the speed of rotation and hence the power output can be controlled, and the blade aerodynamics can be optimized for maximum efficiency. In most cases, the three-blade model is used as it has the highest efficiency in ratio of the number of blades and their overall weight.

At blade design, the actual shapes are very similar within commercial turbines, although, slightly differs by each manufacturer for the best possible aerodynamics according to company preferences [9]. Common characteristics are the hollow design, to reduce weight and the turnable rotor blade tip to help overspeed limitation [10]. Modern blades generally made of Fiber-Reinforces Composites such as carbon fiber and glass fiber with a matrix material of polyester resins or epoxy resins.

Carbon fiber generally has good braking and elasticity characteristics, with stiffness not far from steel, although it is the most expensive material component among the possible choices. Also, in regards to lightning protection, it requires special considerations due to the fact of its material properties which are similar to a semiconductor, creating issues with lightning attachment and flashovers on the surface of the blade.

Glass fiber, on the other hand, has lower ratings in almost all the characteristics mentioned before, but it is considerably cheaper, and it acts as an insulator. Manufacturers tend to use it with more expensive but high-quality epoxy resins to enhance the required physical properties of it [10,11]. Although, the blade is nonconductive, it still attracts lightning due to the fact of its height; therefore, lightning protection is necessary.

3.2. Lightning Protection in General

Lightning protection systems for wind turbines are based on International Electrotechnical Commission (IEC) IEC 61400-24. According to this standard, the lightning protection levels (LPLs) have been set in accordance with the probability of minimum and maximum expected lightning currents, I to IV. The maximum protection, LPL I levels should not be exceeded with a probability of 99% for negative flashes, meanwhile, for positive flashes it is below 10% [12]. The parameters for LPL II and III–IV are the reduced values of LPL I by 75% and 50%, respectively.

The rolling sphere method (RSM) was used to identify the locations of the air termination system on a given structure. The method assumes that there is a spherical region with a radius equal of the striking distance located around the tip of the oncoming lightning leader to a structure. Owing to that, the RSM method demonstrated on a wind turbine with 20 m radius (LPL I). This radius, r, is in relation to the peak current I of the first stroke. According to the IEEE, the equation is:

$$r = 10I^{0.65} \tag{1}$$

There are many different proposals regarding the calculations of the radius for the rolling sphere in relation with the peak current, but the suggested values for each protection level are set by the standards [7] where for each LPL and radius, there is a corresponding minimum peak current value which, against the protection level, it gives protection.

3.3. Protection Methods for Blades

There are four main types of lightning protection methods developed as recommended and outlined in IEC 61400-24 [7]. The methods are as follow:

(a)   receptors placed in the tip and an internal wire (i.e., conductor) is used to carry the current to the hub

(b)   metallic conductor placed around the edges to serve as termination and down conductor

(c)   metal mesh used on the side of the blade

Regardless of the methods, the main function [12–14] of the lightning protection on the blades is:

-   Successful attachment of the lightning strike to a designated or preferred air termination or down conductor system to conduct the current safety without damaging the blades;
-   Provide passage for the lightning current through sufficient cross-section conductors, diverters, and air terminators to earth. Preventing damage to the system and minimizing the high level magnetic and electric field due to high currents;
-   Minimizing the high level of voltages induced and observed inside and outside of the turbine.

With insulator-based materials blades, such as glass fiber composites, the conductors can be placed outside of the blade to divert lightning from the blade surface, also, can be placed inside, with air-terminations at specific point outside of the blade. When carbon fiber composites are used, a layer of conducting material is placed over it which can then carry the current to the blade root. With both cases, sliding connectors are used to carry the current from the blade to the hub towards the ground [12–14].

For the earth termination and down conductors, it has to carry the lightning current safely to the ground where common materials are aluminum, steel, and copper. In general, air termination and for down conductor, the cross-section of at least 50 mm^2 is recommended [7,12–14].

3.4. Lightning Damage to Wind Turbines and Blades

According to many field observations and studies [15–18], wind turbines receive significant amounts of lightning attachments during their designed lifetime, mostly on rotor blades. The damages caused mostly from unsuccessful attachments on air terminations or from induced voltages from electric and electromagnetic fields. The highest percentage of damages occurred on the control system, although, on some cases, the damage were simple interruptions. Meanwhile the damage caused on blades are 11%, it often corresponds with severe damage. The damages associated with lightning are generally blade rupturing and burnout, wire melting, surface cracking and delamination, lightning receptor vaporization, and loss [19–25].

The most popular lightning protection model used nowadays for large turbines consist of an internal down conductor and metal receptors or air terminators penetrating the surface of the blade to serve as desired attachment points. These two systems are then connected together inside of the blade to carry the lightning current to earth. The receptors are installed at nearby the tip of the blade or placed at equal distance from each other alongside the blade from the root to the tip.

One of the main issues with this type of protection is that since the receptors are small compared to the blade planform area, it decreases the efficiency of the attachment of the lightning, causing damage on the surface of the blade [21–25].

Considering the distribution of the lightning attachment and damage along the blade, it can be seen that majority of the attachment occurs at the tip, and the percentage decreases as the distance increasing from the end of the blade. As it can be seen, around 60% of the total damage was located in the last meter of the turbine blade, and 90% of the total damage occurred in the first 4 m [26].

Even though there are many different designs for the lightning protection of blades, there is still potential room for improvement. On the interception of the lightning to the air terminations to increase the effectiveness of the captured lightning flashes and on the down conduction part with the connections of different parts to conduct the current safely to earth.

3.5. Blade Model for Investigation

The blade to be inspected was based on an existing model, currently the largest turbine on the market Vestas V164-9.5MW [9], at present, produced for offshore, although the company is in the process for an onshore model with similar dimensions [27]. For this study, the length of the blades was only considered for the simulation. Although their lightning protection systems are compliant with IEC 61400-24 standards, the exact lightning protection system employed by the blade's manufacturer is not available in the public domain. However, as briefly discussed in Section 3.3, any wind turbine blades should be protected and complied as per methods proposed by IEC 61400-24 standards [7]. For a structure this size, approximately with a tip height around 200 m above sea level, the number of strikes can be estimated considering the lightning density in Europe (between 0.1 and 42 flashes per year per km^2) [5,7].

The height of structure greatly affects the number of flashes predicted on the structure. Based on the regular expected turbine lifetime, what is generally predicted to be 20–25 year, it is very likely that the turbine will be hit at least once during its lifetime. Without any protection, the blade will most likely be destroyed. If the base cost lies between GBP 0.6–0.8 million per MW for an onshore turbine, and generally around 13% of these blades are [28], therefore, the estimated price for losing one blade would be roughly GBP 300,000 on the aforementioned model, not calculating the replacement, transportation, and power outage caused costs. From this, it is clear that wind turbines require adequate protection against lightning strike nevertheless of their location, since even if it is estimated with the lowest density, over the expected lifetime the turbine will be struck at least one or two times.

4. ANSYS Workbench Implementation

4.1. ANSYS Workbench

Nowadays, engineering problems are becoming genuinely complex, relying only on theory, and. physical experiments are not practical anymore. Furthermore, deriving those with hand calculations are rather complex and time consuming. Analysis Systems (ANSYS) is one of the most reputable engineering software analysis packages available on the market and is used by many companies and research facilities around the world. The software is based on finite element analysis (FEA) to solve complex problems in single or multiphysics environment.

The basic principle of the method is that the domain or object is divided into elements with discretization. The distribution of the elements is called mesh, and the points connecting the elements are nodes. When the mesh is generated, an equation is generated for each element regarding with the solvable physics or method of analyzation. The elemental equation is than assembled to a global equation to describe the behavior of the body as a whole [29].

4.2. Blade and Protection Implementation

As briefly discussed in Sections 2 and 3, the wind turbine is a grounded structure, hence, the lightning current as a result of return stroke will then be passed safely to the ground through hub, nacelle, tower, and tower footing at the ground level. Hence, when lightning strike on the lightning receptor installed on a blade, the ground is elevated to the highest tip at the time of strike due to the blade tip being at its highest point at a time. Thus, this assumption is also used by many other lightning researchers around the world [1,3,4,8,10,14,15,19,20,26] and also for this study. Owing to that, single blade was examined without any attachment to rotor and nacelle. The ANSYS Workbench version 18.2 was used to carry out the simulation of the lightning protection of blade. The available software license was for Academic Research, which restricts the meshing node number to 300,000 which corresponds to around 40,000 elements depending on the meshing algorithm chosen. As it was mentioned earlier, the size base was taken from an existing model (Vestas V164-9.5 MW. The turbine had approximately 80 m long blades; in the model it was extended slightly to represent a potential future size. As shown in Figure 1, the hollow blade design can be seen from what was modelled in ANSYS DesignModeler. The model measurements were 85 m long, 5 m wide, 2.6 m depth, 10 times base-to-tip ratio, and 0.015 m wall thickness. The current was applied at point A, meanwhile, points B and C were specified as 0 V.

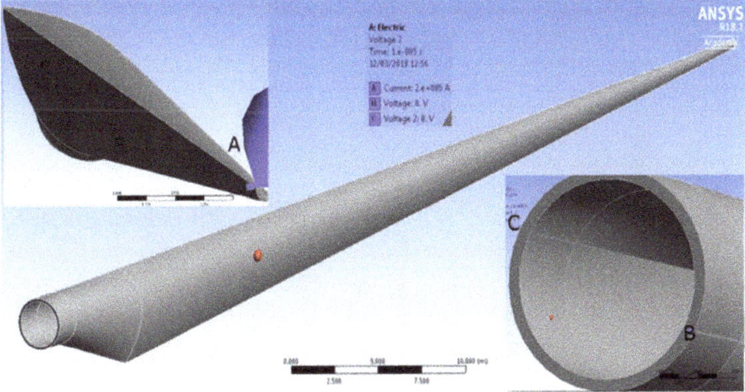

Figure 1. Wind turbine blade for simulation.

The blade material was chosen to be E-glass fiber-reinforced polyester with the necessary values set manually [30,31] to serve as an insulator-type blade, the lightning conductor was set to the copper parameters taken from standards [12] with a 50 mm^2 round cross-section as the minimal specified area. For evaluation, one of the recommended method by IEC 61400-24 [7] was considered where this method was used previously for smaller turbines, although in this project, it was examined for larger turbine blade.

4.3. Simulation Setup

For the lightning attachment point, a part on the conductor at the tip of the blade was defined (point A). In the absence of specifying ground, 0 voltage was applied on the connecting ends of the conductor (points B and C (Figure 1)).

For simulation, electric, transient-thermal, and static structural analysis was chosen using Mechanical APDL solver [32]. As shown in Figure 2, the applied mechanical APDL structure can be seen. By connecting the electric, thermal, and structural sections, it was possible to transfer results from one stage to another, creating a complex simulation environment.

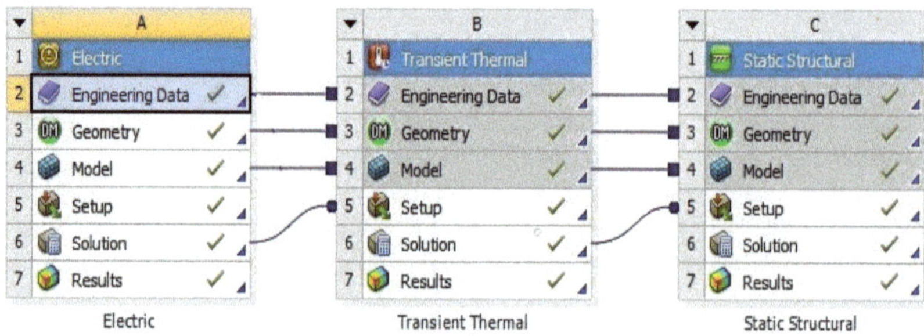

Figure 2. Simulation setup for the model.

The first and subsequent return stroke current rise were implemented according to the current standards [12], with an additional 'extreme' level of first and subsequent return stroke and the effects were observed over set amount of time as tabulated in Table 1. The 'extreme' level used referred to the highest recorded lightning peak current [2,5].

Table 1. Test parameters for simulations showing the extreme case for LPL [23].

Type of Stroke	Test Parameter	Unit	LPL
			Extreme (0)
First	Δi	kA	300
	Δt	μs	10
Subsequent	Δi	kA	75
	Δt	μs	0.25

The ambient temperature was set to 20 °C, and the blade was set to be fixed at the base. For testing the proposed method, first, the cross-section of the down conductor area was set to the recommended minimum area which was then increased to 100 mm^2 and to 200 mm^2. Afterwards, as it has been mentioned in many publications [4,26,33–35] and stated in the standards [7], lightning tends to attach to the tip and to the close approximation of the blades. Therefore, to overcome the destructive heating effect of the lightning, especially at the attachment point on the conductor, a hybrid conductor has been designed. This design consisted of two conductors with different diameters joined together. The larger diameter covered the tip of the blade and ran down at a specific distance from the tip towards the root. The joints of the two conductors could be welded or the whole conductor could be molded to achieve a better transition between the different thicknesses. In total, six case studies were examined with different diameters and a combination of conventional and hybrid methods:

<u>**Conventional:**</u>

A: Minimal protection level with 50 mm^2 conductor cross-section area;

B: 100 mm^2 conductor cross-section area;

C: 200 mm^2 conductor cross-section area.

<u>**Hybrid:**</u>

D: Hybrid conductor design for tip;

E: Hybrid conductor design, 2 m on sides;

F: Hybrid conductor design, 5 m on sides.

The parameters examined from the simulation models were:

1. Voltage at the attachment point (V);
2. Maximum value of Joule heating in the conductor (MW/m^3);
3. Current density at the attachment point (kA/m^2);
4. Maximum temperature generated by Joule heating in the conductor (°C);
5. Total deformation caused on the blade due to the heating effect (mm);

In addition to the existing parameters, another probe was added to hybrid design at the joints of the two conductors to follow the change in the current density:

6. Current density at joint (kA/m^2).

5. Simulation Results and Discussion

5.1. Results from Conventional Case Studies

The graphical representation of the conventional design can be seen in Figure 3, where:

- Current density in the conductor;
- Temperature generated by current;
- Deformation caused by temperature.

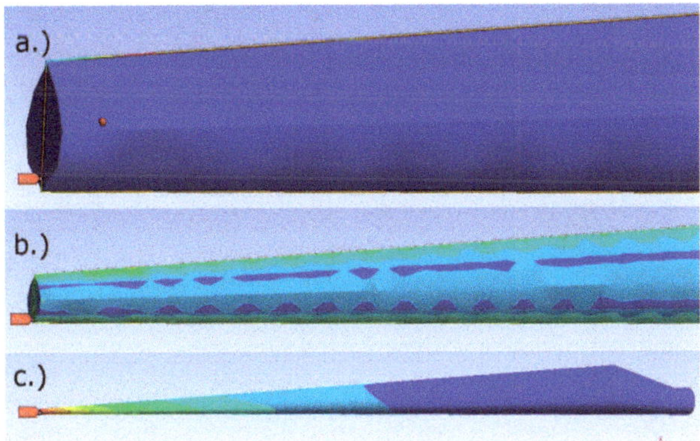

Figure 3. Graphical simulation results for conventional design; (a) 50 mm^2 cross section area conductor, (b) 100 mm^2 cross section area conductor; (c) 200 mm^2 cross section area conductor.

As the current from the lightning strike runs from the striking point towards ground (0 V), it heats up the conductor and the blade at the contact surfaces. Due to thermal expansion, the blade and the conductor experience force which causes deformation in both bodies.

5.1.1. Case Study A: Minimal Protection Level with 50 mm^2 Conductor Cross-Section Area

For first case, the conductor cross-section area was set to the minimal recommended 50 mm^2 value where the results of first and subsequent strokes are tabulated in Table 2. Joule heating or resistive heating occurs when electric current passing through a conductor with resistance and it is proportional to the resistance of the conductor and square of the current [36]. The maximum Joule heating occurs close to the attachment point; therefore, the maximum temperature appears at the exact same location. Thus, the highest total deformation can be seen around the tip, where the current enters the conductor.

Table 2. Results obtained from the first and subsequent return strokes for **Case Study A.**

First Strokes	LPL 0 300 kA	LPL I 200 kA	LPL II 150 kA	LPL III-IV 100 kA
Voltage at striking point (V)	4360.3	2969.9	2180.2	1453.4
Maximum joule heating (MW/m^3)	191,420	85,070	47,855	21,269
Current density at striking point (kA/m^2)	823,240	548,830	411,620	274,410
Maximum temperature (°C)	608.98	282.88	168.74	87.22
Total deformation (mm)	47.725	2.1232	1.196	5.337
Subsequent Strokes	**LPL 0 75 kA**	**LPL I 50 kA**	**LPL II 37.5 kA**	**LPL III-IV 25 kA**
Voltage at striking point (V)	1090.1	726.72	545.04	363.36
Maximum joule heating (MW/m^3)	11,964	5317.2	2991	1329.3
Current density at striking point (kA/m^2)	205,810	137,210	102,910	68,604
Maximum temperature (°C)	58.686	38.305	31.172	26.076
Total deformation (mm)	0.3019	0.13632	0.078379	0.037002

5.1.2. Case Study B: 100 mm^2 Conductor Cross-Section Area

For this case, the conductor cross-section area was increased to 100 mm^2 where the results are tabulated in Table 3 and it shown that the same parameters observed in Case Study A were decreasing as the cross-section area increases. This has been anticipated as the conductor cross-section increases, it decreases the resistance, therefore the heating and deformation as well.

Table 3. Results obtained from the first and subsequent return for **Case Study B.**

First Strokes	LPL0 300 kA	LPL I 200 kA	LPL II 150 kA	LPL III-IV 100 kA
Voltage at striking point (V)	1937.9	1292	968.97	645.98
Maximum joule heating (MW/m^3)	39,689	17,639	9922.1	4409.8
Current density at striking point (kA/m^2)	407,620	271,700	203,810	135,870
Maximum temperature (°C)	149.31	78.583	53.828	36.146
Total deformation (mm)	0.79158	0.35234	0.19862	0.08883
Subsequent Strokes	**LPL 0 75 kA**	**LPL I 50 kA**	**LPL II 37.5 kA**	**LPL III-IV 25 kA**
Voltage at striking point (V)	484.49	322.99	242.24	161.5
Maximum joule heating (MW/m^3)	2480.5	1102.5	620.13	276.1
Current density at striking point (kA/m^2)	101,900	68,936	50,952	33,968
Maximum temperature (°C)	29.957	25.536	23.989	23.081
Total deformation (mm)	0.0504	0.02311	0.01364	0.00714

5.1.3. Case Study C: 200 mm^2 Conductor Cross-Section Area

For this case, the cross-section of the lightning conductor was further increased to 200 mm^2. As was expected, the resulting values further decreased as the cross-section increased (Table 4).

Table 4. Results obtained from the first and subsequent return strokes for **Case Study C**.

First Strokes	LPL0 300 kA	LPL I 200 kA	LPL II 150 kA	LPL III-IV 100 kA
Voltage at striking point (V)	1090	726.66	545	363.33
Maximum joule heating (MW/m^3)	13,389	5950.8	3347.3	1487.7
Current density at striking point (kA/m^2)	302,270	201,510	151,130	100,760
Maximum temperature (°C)	65.017	41.118	32.754	26.78
Total deformation (mm)	0.27567	0.13922	0.079422	0.036739
Subsequent strokes	LPL 0 75 kA	LPL I 50 kA	LPL II 37.5 kA	LPL III-IV 25 kA
Voltage at striking point (V)	181.67	136.25	90.833	181.67
Maximum joule heating (MW/m^3)	371.93	209.21	92.982	371.93
Current density at striking point (kA/m^2)	50,378	37,784	25,189	50,378
Maximum temperature (°C)	23.195	23.08	23.069	23.195
Total deformation (mm)	0.01129	0.00769	0.00523	0.01129

5.2. Results from Hybrid Case Studies

Observing the previously acquired results, the conductor design was constructed to decrease the effects caused by the lightning stroke. By increasing the diameter of the conductor at the tip of the blade, should decrease the impact of the current on the body. The following three cases have been developed and examined with different length of the increased area conductor as depicted in Figure 4.

Figure 4. Hybrid conductor designs; (**a**) thicker conductor for tip, (**b**) thicker conductor for tip with 2 m on sides, (**c**) thicker conductor for tip with 5 m on sides.

5.2.1. Case Study D: Hybrid Conductor Design for Tip

In this case, the conductor cross-section area at the tip of the blade has been increased to 100 mm^2, meanwhile the rest of the conductor has been left at the minimum recommended value. Assuming that the change in diameter of the attachment area, the impact of the lightning strike attached on the conductor reduced as anticipated where results as tabulated in Table 5.

5.2.2. Case Study E: Hybrid Conductor Design, 2 m Sides

In the second case for the hybrid strategy, the length of the thicker conductor increased for 2 m on the sides of the blade, this hypothetically should further decrease the effects caused by the lightning strike. The results are tabulated in Table 6.

Table 5. Results obtained from the first and subsequent return strokes for **Case Study D.**

First Strokes	LPL0 300 kA	LPL I 200 kA	LPL II 150 kA	LPL III-IV 100 kA
Voltage at striking point (V)	4354.3	2902.8	2177.7	1451.4
Maximum joule heating (MW/m^3)	341,640	151,840	85,411	37,961
Current density at striking point (kA/m^2)	1,091,100	727,370	545,530	363,690
Maximum temperature (°C)	547.52	255.56	153.38	80.391
Total deformation (mm)	4.5937	2.043	1.502	0.5125
Current density at joints (kA/m^2)	2,715,500	1,810,400	1,357,800	905,180
Subsequent Strokes	**LPL 0** 75 kA	**LPL I** 50 kA	**LPL II** 37.5 kA	**LPL III-IV** 25 kA
Voltage at striking point (V)	1088.6	725.71	544.28	362.85
Maximum joule heating (MW/m^3)	21,353	9490.1	5338.2	2372.5
Current density at striking point (kA/m^2)	272,760	181,840	136,380	90,922
Maximum temperature (°C)	54.845	36.598	30.211	25.649
Total deformation (mm)	0.2893	0.1299	0.0741	0.0343
Current density at joints (kA/m^2)	678,890	452,590	339,440	263,000

Table 6. Results obtained from the first and subsequent return strokes for **Case Study E.**

First Strokes	LPL0 300 kA	LPL I 200 kA	LPL II 150 kA	LPL III-IV 100 kA
Voltage at striking point (V)	2773.4	1848.9	1386.7	924.45
Maximum joule heating (MW/m^3)	66,067	29,363	16,517	7340.8
Current density at striking point (kA/m^2)	844,930	563,290	422,470	281,640
Maximum temperature (°C)	237.15	117.62	75.788	45.906
Total deformation (mm)	1.355	0.6034	0.3404	0.1525
Current density at joints (kA/m^2)	2,072,100	1,381,400	1,036,100	690,700
Subsequent Strokes	**LPL 0** 75 kA	**LPL I** 50 kA	**LPL II** 37.5 kA	**LPL III-IV** 25 kA
Voltage at striking point (V)	693.34	462.23	346.67	23.11
Maximum joule heating (MW/m^3)	4129.2	1631.3	1032.3	458.8
Current density at striking point (kA/m^2)	211,230	140,820	105,620	70,411
Maximum temperature (°C)	35.447	27.978	25.362	23.494
Total deformation (mm)	0.0867	0.0398	0.0234	0.0118
Current density at joints (kA/m^2)	518,030	345,350	259,010	172,680

5.2.3. Case Study F: Hybrid Conductor Design, 5 m Sides

The thicker portion of the conductor has been further increased to 5 m on each side of the tip of the blade, in theory, further reducing the recorded values where results are tabulated in Table 7.

Table 7. Results obtained from the first and subsequent return strokes for **Case Study F.**

First Strokes	LPL0 300 kA	LPL I 200 kA	LPL II 150 kA	LPL III-IV 100 kA
Voltage at striking point (V)	2691.2	1794.1	1345.6	897.06
Maximum joule heating (MW/m^3)	64,236	28,549	16,059	7137.3
Current density at striking point (kA/m^2)	495,610	330,410	247,800	165,200
Maximum temperature (°C)	218.42	115.85	74.788	45.461
Total deformation (mm)	1.216	0.5420	0.3061	0.1376
Current density at joints (kA/m^2)	1,999,800	1,333,200	999,910	666,610
Subsequent Strokes	**LPL 0** 75 kA	**LPL I** 50 kA	**LPL II** 37.5 kA	**LPL III-IV** 25 kA
Voltage at striking point (V)	672.8	488.53	336.4	224.27
Maximum joule heating (MW/m^3)	4014.7	1784.3	1003.7	446.08
Current density at striking point (kA/m^2)	123,900	82,601	61,951	41,301
Maximum temperature (°C)	35.197	27.865	25.299	23.466
Total deformation (mm)	0.0787	0.0366	0.0218	0.0114
Current density at joints (kA/m^2)	499,960	333,300	249,980	166,650

5.3. Discussion

5.3.1. Conventional Cases (A, B, and C)

Based on the simulation results on the conventional type conductor tabulated in Tables 2–4, the current density at the attachment point and total deformation (highest value at the tip of the blade) plot can be seen on Figure 5. Comparing the graphs, it can be seen that increasing the diameter of the conductor reduces the value of current density and the amount of deformation produced on the blade as it can be expected. Although, further inspecting the results, the difference between the 100 mm^2 and the 200 mm^2 cross-section area was less significant (34.8%) than the difference between the 50 mm^2 and 100 mm^2 (101.9%). Increasing the diameter of the conductor implied better results or lower values, although it was not linear compared to the change in diameter. Evaluating the results leads to the assumption that the 100 mm^2 cross-section area produced the best results among the tested values according to the given LPL, considering the weight and cost of the usable material. Similar correlations can be seen on the graphs from the rest of the results in Figure A1 (Appendix A).

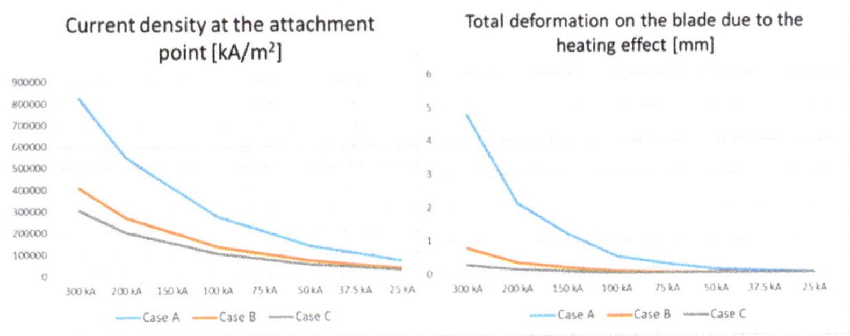

Figure 5. Current density (**left**) and total deformation of the blade (**right**) for the conventional case studies.

5.3.2. Hybrid Cases (D, E, and F)

As shown in Figure 6, the graphical representation of the simulation results can be seen for Case E.

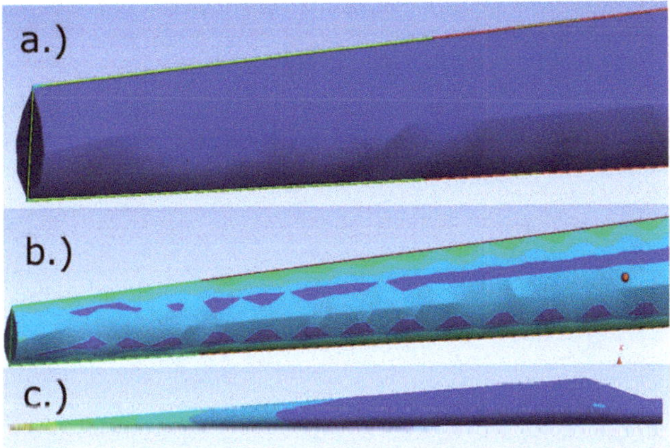

Figure 6. Graphical simulation results, hybrid design (case study E); (**a**) current density in the conductor, (**b**) temperature generated by current, (**c**) deformation caused by temperature.

Based on the data tabulated in Tables 5–7, Figure 7 plotted the different current densities at different points comparing three different hybrid cases. When diameters increased, the current density reduced in the conductor as well as heating and deformation in the blade. On the other hand, at the joints of the two types of conductor, there was still an increment that could still be seen. Comparing the values of the three designs indicates that the tip only version reduces the effects of the stroke at the attachment point the least, although increasing the length of the higher diameter conductor reduces the current density at both the attachment point and at joint. The increase in current density between the attachment point and the joint, for case D, case E, and case F with 148.9%, 145.2%, and 303.5%, respectively, thus the possible damage due to the current flowing in the down conductors. This suggests that the most efficient way to improve the LPS would be to increase the overall diameter of the whole conductor.

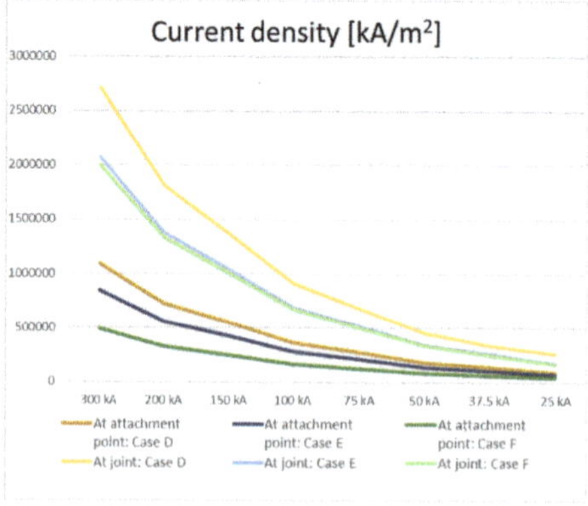

Figure 7. Current density of hybrid case studies.

As shown in Figure 8, on the left, the maximum temperature of the blade, which was measured at conductor joints and on the right the total deformation caused on the blade, where the highest measured value was at the tip of the blade. It shows that there was no significant reduction in temperature and deformation between the 2 m and 5 m type (8% for temperature and 11.4% for deformation), although, the maximum temperature point moved from the area of attachment point to the joints of the two conductor. Comparing the three tested designs' results, the 2 m long conductor is suggested to be the most sufficient of all, considering the amount of material involved and the improvement in temperature, thus reduction in deformation too. The rest of the results can be seen in Figure A2.

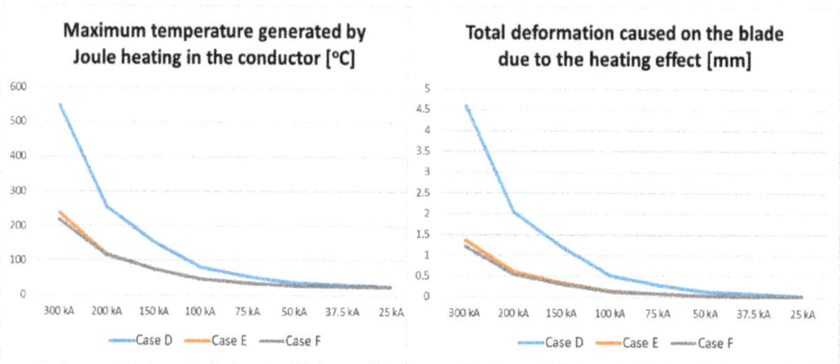

Figure 8. Maximum temperature (**left**) and total deformation of the blade (**right**) for the hybrid case studies.

5.3.3. Conventional and Hybrid

As shown in Figure 9, on the left, the maximum temperature of the blade, meanwhile on the right, the total deformation caused on the blade, where the highest measured value was at the tip of the blade. It can be seen that both the B and E cases performed better compared to the minimal conductor cross-section area in terms of temperature increment and blade deformation. For LPL 0, the temperature difference between Case B and A was 459.67 °C (307.86%), meanwhile between Case E and A it was 371.83 °C (156.79%), furthermore, the temperature difference between Case B and E is 87.74 °C. Furthermore, the temperature increase is linearly proportional to the deformation and the changes for deformation are nearly identical.

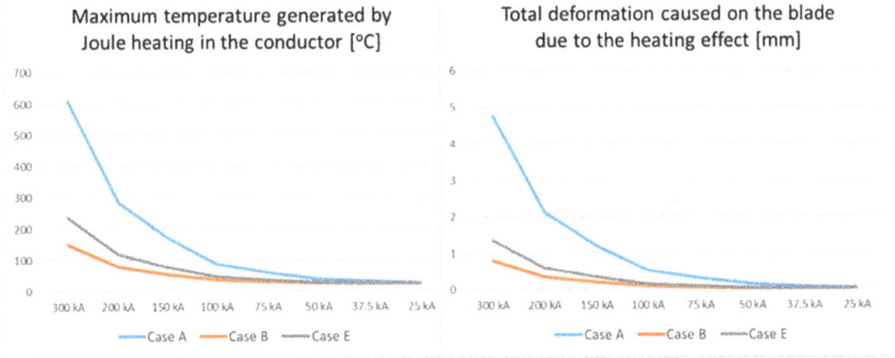

Figure 9. Maximum temperature (**left**) and total deformation (**right**) of the blade at A, B, and E case studies.

Comparing Cases B and E, the results showed that Case B suggests the most effective conductor design in terms of temperature and displacement, also the other measured parameters.

5.3.4. Summary

Comparing results obtained from Case Studies, it can be deduced that the most efficient way of increasing the efficiency of the protection of a wind turbine blade is to increase the diameter of the down conductor. However, it will be compromised due to the actual cost of the extra material, the weight increasement and the possible effects on the airfoil of the blades as these factors are the most crucial in blade design. Furthermore, applying Case E to a modern turbine blade could potentially reduce the effects of the heat and deformation because it only requires small portion of the conductor in the tip region. In general, most turbines are glued together at the leading and trailing edge to reduce the manufacturing costs, this brings an issue since the down conductors placed along these lines. As the current heats up the conductor, this could possibly melt the applied glue material causing severe damage which may cause blade to separate. As to potentially alleviate this, it would be possible to implement the hybrid conductor design for the blade lightning protection.

6. Conclusions, Recommendations for Future Works

6.1. Conclusions

Lightning protection is an important aspect of wind energy, since over the expected lifetime of a turbine; at least once a lightning will hit it. Due to the enormous amount of current, without proper protection, it is most likely to result failure to turbine and will cause high repair costs. The protection methods and levels are proposed by standards to achieve the minimal protection suggested, although this protection cannot be taken as guarantee for all cases. As wind turbines keep increase in size to keep up with the generation demand, as the chance increases, of being hit by a lightning due to their elevation from ground level.

In this paper, a conventional lightning protection concept, previously used for smaller turbine models has been evaluated for possible use for large blades. For the task, simulation software, ANSYS Workbench, Mechanical APDL has been used. In the first three case studies, different conductor cross-section areas have been set for conventional design for full length of the conductor. For the second half of the case studies a hybrid conductor model was evaluated. This design consists of two conductors with different diameters joined together. The higher diameter one covered the tip of the blade and ran down at specific distance from the tip towards the root. The lightning parameters was set according to the current standards, with and additional extreme first and subsequent return stroke current amplitude. Comparing the simulation outcomes has been showed that Case Study C indicated the most promising results among all. In the other hand considering the weight and cost of the extra material, also the possible aerodynamical effects of the conductor around the blade, Case Study E has been appeared to be the most adequate alternative. The design shows great improvement in reducing the lightning caused effects, compared to Case Study A, therefore the possible damage on the blade. Furthermore, it only requires simple modification of the existing lightning protection concept, minimizing the associated costs, weight, and the possible disturbance in the aerodynamics of the blade.

6.2. Recommendations for Future Work

There are still many factors and values that should be evaluated in order to give full understanding and clarification of the proposed design.

- One possible future work could include the examination of electromagnetic forces and waves generated by the current, since those were excluded from the simulation due to missing Mechanical APDL functions (electromagnetic analysis system).
- There are possible incorrect, unrealistic values presented in this work due to the potential misconfiguration of simulation physics in the absence of relatable guide.
- The software used had limited solver size due to academic license; therefore, the mesh of the objects had to be left coarse, meaning less accurate and possible differences in expected and real life values.
- The design and therefore the investigation could be extended to model a complete turbine to see the effects on the whole structure.
- The exact length of the increased diameter conductor could be evaluated in the ratio of the size of the blade; therefore, the proposed design could be implemented on various size blades with maximum efficiency.
- A comparison could be made with existing blade LPS what is used on large blades nowadays to estimate the efficiency of both designs.

Author Contributions: Conceptualization, A.S.A.; Data curation, A.S.A.; Formal analysis, V.M.; Funding acquisition, F.M.-S., A.S.M.S. and J.A.A.-R.; Investigation, V.M. and A.S.A.; Methodology, A.S.A.; Resources, A.S.A.; Software, A.S.A.; Supervision, A.S.A.; Validation, A.S.A.; Visualization, A.S.A. and F.M.-S.; Writing—original draft, V.M.; Writing—review and editing, V.M., A.S.A., F.M.-S., M.Z., M.N.M., A.S.M.S. and J.A.A.-R. All authors have read and agreed to the published version of the manuscript.

Funding: Part of the work presented in this research study funded by the Agencia Nacional de Investigación y Desarrollo (through the project Fondecyt regular 1200055 and the project Fondef ID19I10165), project PI_m_19_01 (UTFSM) and by Universiti Kuala Lumpur under the Short Term Research Grant (STRG) STR18022.

Conflicts of Interest: The authors declare no conflict of interest.

Appendix A

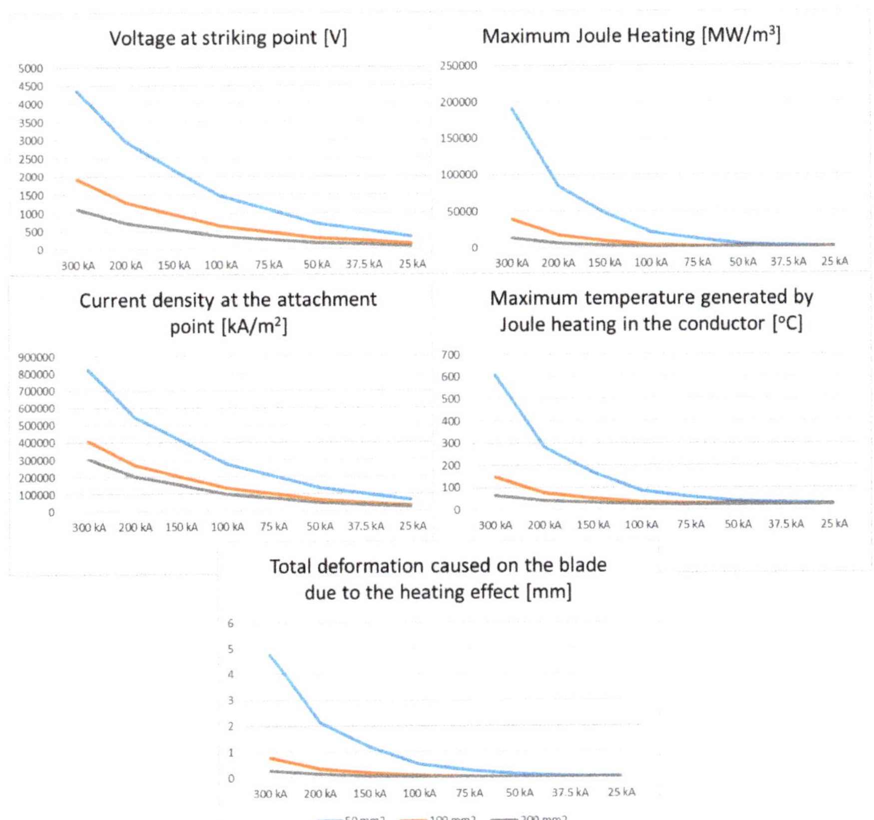

Figure A1. Plotted results from conventional simulations.

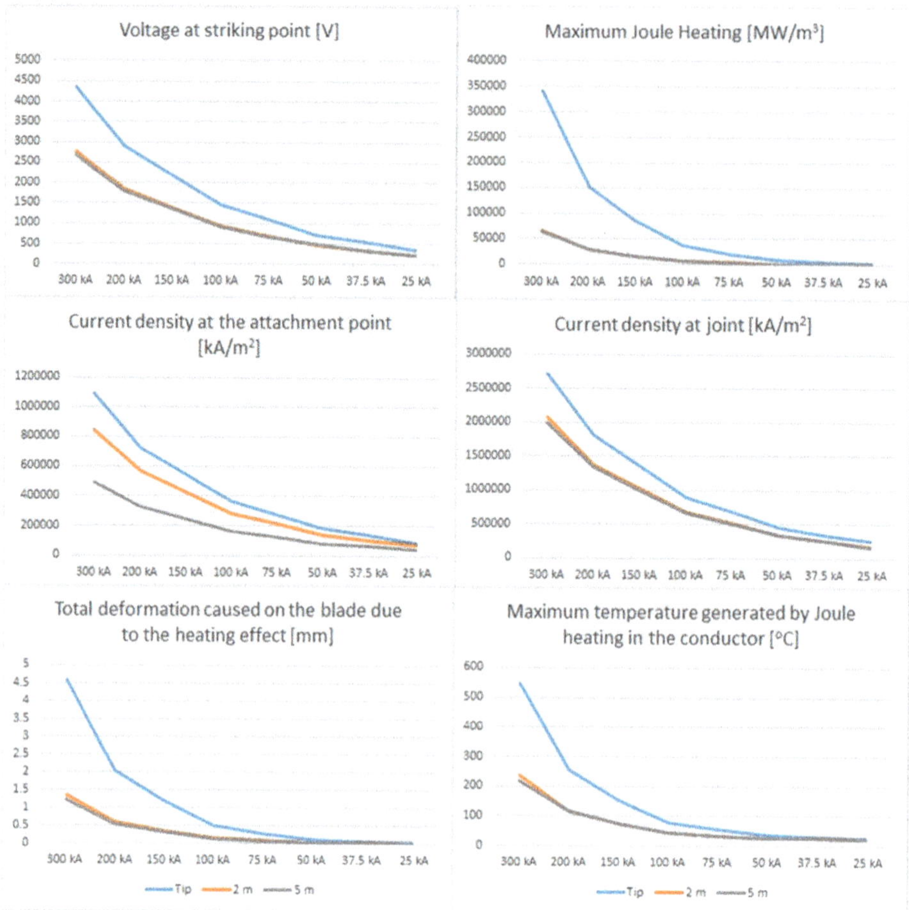

Figure A2. Plotted results from hybrid simulations.

References

1. Cooray, V. *An Introduction to Lightning*; Springer: Heidelberg, Germany, 2015.
2. NSSDC. Sun Fact Sheet. Available online: https://nssdc.gsfc.nasa.gov/planetary/factsheet/sunfact.html (accessed on 2 October 2018).
3. Rakov, V.A.; Uman, M.A. *Lightning: Physics and Effects*; Cambridge University Press: Cambridge, UK, 2003.
4. Cooray, V. *Lightning Protection*; The Institution of Engineering and Technology: London, UK, 2010.
5. The Weather Channel. Where Almost 9 Billion Lightning Strikes in 5 Years Have Happened on Earth. Available online: https://weather.com/safety/thunderstorms/news/2018-05-07-world-lightning-strikes-map-vaisala-2013-2017 (accessed on 2 February 2019).
6. Anderson, G.; Klugmann, D. European lightning density analysis using 5 years of ATDnet data. *Nat. Hazards Earth Syst. Sci.* **2014**, *14*, 815–829. [CrossRef]
7. Technical Committee PEL/88. *Wind Turbines Part. 24: Lightning Protection, BS EN 61400-24:2019*; The British Standards Institution: London, UK, 2019.
8. Uman, M.A. *The Lightning Discharge*; Dover Publications, Inc.: Mineola, NY, USA, 2001.

9. HMI Vestas Offshore Wind. The World's Most Powerful Available Wind Turbine Gets Major Power Boost. Available online: http://www.mhivestasoffshore.com/worlds-most-powerful-available-wind-turbine-gets-major-power-boost/ (accessed on 5 February 2019).

10. Hu, W. *Advanced Turbine Wind Technology*; Springer: Ithaca, NY, USA, 2018.

11. Czeshop. Wind Turbine Blade Cross Section. Available online: http://czeshop.info/2017/wind-turbine-blade-cross-section.html (accessed on 10 February 2019).

12. Technical Committee GEL/81. *Protection against Lightning Part. 1: General Principles, BS EN 62305-1:2011*; The British Standards Institution: London, UK, 2017.

13. Cooray, V. *Lightning Protection*; Athenaeum Press Ltd.: Gateshead, UK, 2010.

14. Barwise, A. Lightning and Surge Protection for Wind Turbines. Available online: https://www.ee.co.za/article/lightning-surge-protection-wind-turbines-2.html (accessed on 21 February 2019).

15. Wilson, N.; Myers, J.; Cummins, K.L.; Hutchinson, M.; Nag, A. Lightning attachment to wind turbines in Central Kansas: Video observations, correlation with the NLDN and in-situ peak current measurements. In Proceedings of the European Wind Energy Conference Exhibition EWEC 2013, Vienna, Austria, 4–7 February 2013.

16. Cummins, K.L.; Zhang, D.; Quick, M.G.; Garolera, A.C.; Myers, J. Overview of the Kansas Windfarm2013 Field Program. In Proceedings of the 23rd International Lightning Detection Conference (ILDC), Tucson, AZ, USA, 18–19 March 2014.

17. Wang, D.; Takagi, N.; Watanabe, T.; Sakurano, H.; Hashimoto, M. Observed characteristics of upward leaders that are initiated from a windmill and its lighting protection tower. *Geophys. Res. Lett.* **2008**, *35*. [CrossRef]

18. Ishii, M.; Saito, M.; Natsuno, D.; Sugita, A. Lighting current observed at wind turbines at winter in Japan. In Proceedings of the International Conference on Lightning & Static Electricity ICLOSE 2013, Seattle, WA, USA, 18–20 September 2013.

19. Yasuda, Y.; Yokoyama, S.; Minowa, M. Classification of Lightning Damage to Wind Turbine Blades. *IEEJ Trans. Electr. Electr. Eng.* **2012**, *7*, 559–566. [CrossRef]

20. Sorensen, T.; Jensen, F.V.; Raben, N. Lightning protection for offshore wind turbines. In Proceedings of the 16th International Conference & Exhibition on Electricity Distribution, Amsterdam, The Netherlands, 18–21 June 2001.

21. Solacity. Lightning Protection. Available online: https://www.solacity.com/lightning-protection/ (accessed on 30 March 2019).

22. Yasuda, Y.; Yokoyama, S. Proposal of Lightning Damage Classification to Wind Turbine Blades. In Proceedings of the 7th Asia Pacific International Conference on Lightning (APL), Chengdu, China, 1–4 November 2011; pp. 368–371.

23. GlobalBladeService. Lightning Damage. Available online: https://www.globalbladeservice.eu/blade-service/lightning-damage (accessed on 31 March 2019).

24. Gorgeouswithattitude. 1st Blade Wends Way to Wind Farm. Available online: http://gorgeouswithattitude.blogspot.com/2010/09/1st-blade-wends-way-to-wind-farm.html (accessed on 31 March 2019).

25. EPSRC, Lightning Protection of Wind Turbine. Available online: https://community.dur.ac.uk/supergen.wind/docs/presentations/2010-04-25_1425_EWEC%202010_SideEvent_Lightning_VP.pdf (accessed on 31 March 2019).

26. Garolera, A.C.; Madsen, S.F.; Nissim, M.; Myers, J.D.; Holboell, J. Lightning Damage to Wind Turbine Blades From Wind Farms in the U.S. *IEEE Trans. Power Deliv.* **2016**, *31*, 1043–1049. [CrossRef]

27. MHI Vestas Offshore Wind A/S. V164-9.5 MW. Available online: http://www.mhivestasoffshore.com/category/v164-9-5-mw/ (accessed on 24 February 2019).

28. Schubel, P.J. Technical cost modelling for a generic 45-m wind turbine blade producedby vacuum infusion (VI). *Renew. Energy* **2010**, *35*, 183–189. [CrossRef]

29. Alawadhi, E.M. *Finite Element Simulations Using ANSYS*, 2nd ed.; CRC Press: Boca Raton, FL, USA, 2015.

30. AZoNetwork. E-Glass Fibre. Available online: https://www.azom.com/properties.aspx?ArticleID=764 (accessed on 25 February 2019).

31. MatWeb. Glass Fiber Summary List. Available online: http://www.matweb.com/search/DataSheet.aspx?MatGUID=5c6df0ae272e41808562d3374d7b4f5a&ckck=1 (accessed on 28 February 2019).

32. ANSYS, Inc. Analysis Types. Available online: https://www.sharcnet.ca/Software/Ansys/18.2.2/en-us/help/wb_sim/ds_simulation_types.html (accessed on 2 March 2019).

33. Heidler, F.; Flisowski, Z.; Zischank, W.; Bouquegneau, C. Parameters of lightning current given in IEC 62305–Background, experience and outlook. In Proceedings of the 29th International Conference on Lightning Protection ICLP 2008, Uppsala, Sweden, 23–26 June 2008.
34. Guo, Z.; Li, Q.; Ma, Y. Experimental study on lightning attachment manner to wind turbine blades with lightning protection system. *IEEE Trans. Plasma Sci.* **2019**, *47*, 635–646. [CrossRef]
35. Arif, W.; Li, Q.; Guo, Z.; Ellahi, M.; Wang, G.; Siew, W.H. Experimental study on lightning discharge attachment to the modern wind turbine blade with lightning protection system. In Proceedings of the 13th International Conference on Emerging Technologies (ICET), Islamabad, Pakistan, 27–28 December 2017; pp. 1–6.
36. Young, J.H. Steady state Joule heating with temperature dependent conductivities. *Appl. Sci. Res.* **1986**, *43*, 55–65. [CrossRef]

MDPI

St. Alban-Anlage 66

4052 Basel

Switzerland

Tel. +41 61 683 77 34

Fax +41 61 302 89 18

www.mdpi.com

Applied Sciences Editorial Office

E-mail: applsci@mdpi.com

www.mdpi.com/journal/applsci